Contents

Section 1: The enquiry process

Section 2: Methodologies

Section 3: Concepts

Insight and Perspective Ltd, 701 Stonehouse Park, Sperry Way,
Stonehouse, Glos, GL10 3UT

www.insightandperspective.co.uk

First published 2018
10 9 8 7 6 5 4 3 2 1
ISBN 13: 978 1 912190 02 7

Designed and typeset by Wooden Ark
Printed by TJ International, Padstow, Cornwall, UK

Acknowledgements

The author and publishers would like to thank Christine Wise and Andy Leeder for their contributions and input into this Handbook.

The author and publishers would like to thank the following for permission to use the photographs/copyright material:

p43 Hydrograph www.riverlevels.uk/esk, front cover, pp58, 59 Field Studies Council, pp91, 108, 115, 120 Andy Leeder, p110 Environment Agency.

p112 a & b OS Mapping of Felixstowe © Crown copyright 2018
OS 100059646

All other photographs Andy Owen

The publishers have made every effort to trace the copyright holders. If they have inadvertently overlooked any they will be pleased to acknowledge these at the first available opportunity.

What is the enquiry process?

Every good piece of geography fieldwork goes through six stages of an enquiry process. You begin by deciding on your **Aim**. This will relate to one of geography's **concepts** – big ideas that help us understand how the world works. These concepts are listed in Figure 1.

Figure 1 The six concepts that are assessed in GCSE fieldwork

Concept	What is this concept about?	Page
Place	Each place has features that give that place a unique character. People recognise this character – it gives the place, and the people who live there, a sense of identity.	90
Spheres of influence	Geographical features have positive and negative effects on the environment surrounding them. This area around a feature is called its sphere of influence. The nature of the feature influences the size and impact of the sphere of influence.	96
Cycles and flows	Environments contain flows, for example, water, people, and money. These move between stores. This concept explores the factors that influence the size and rate of these flows.	102
Mitigating risk	All environments contain risks. These risks can be reduced (or mitigated) by managing the hazard or by changing the behaviour of people who they may affect.	108
Sustainability	Sustainability is a measure of how well a place or environment is coping with social, economic, or environmental change.	114
Inequality	Features are rarely distributed evenly through the environment. Inequality occurs where one place, or one group of people, has more than its fair share of something.	120

The actual fieldwork is then about collecting the **Evidence** (or data) that will help you to answer your aim. This might be data that you collect yourself (called **primary data**) or information you find somewhere else, for example, on a website (called **secondary data**).

Your data will then need to be processed. This may involve some number crunching and then creating graphs and maybe a map. This will help you to analyse the data and reach a conclusion. Have you found the evidence that proves or disproves your aim? The enquiry process is shown in Figure 2. Each stage of the enquiry process is described fully in the six chapters in Section 1 of this book.

How to use this book

This book has been written to help you enjoy your fieldwork and help you understand what you are doing so that you don't have to get stressed about the examination.

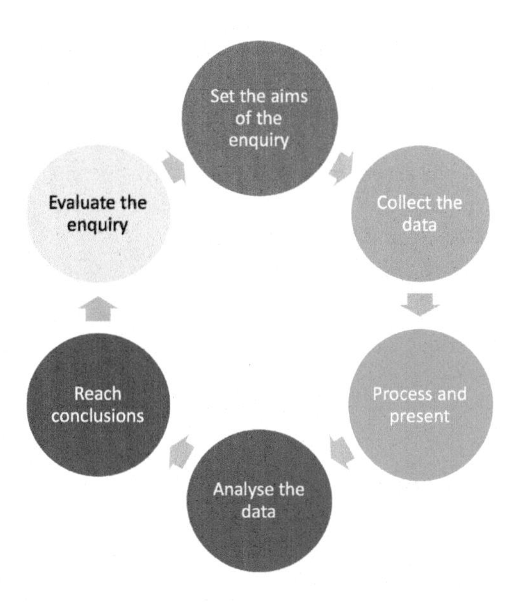

Figure 2 The six stages of the fieldwork enquiry process

Section 1 gives you advice on how to approach each stage of the enquiry process. Understanding each of these stages is vital if you want to do well in your fieldwork.

- Use Chapters 1 and 2 before your fieldtrip to help you plan an aim for your fieldwork enquiry and decide how to collect your data.
- Use Chapters 3 to 6 after the fieldtrip. These chapters will help you to select and draw the most suitable maps and graphs. They will also help you to analyse your evidence, reach a conclusion, and evaluate what you have done.
- Use Section 1 of the book again when you are revising for Component 3.

Section 2 describes details of four different methods of collecting data. You can use more than one of these methods to collect data for your fieldwork enquiry. For example, you might decide to collect data along a **transect** (which means to sample along a line). The data you collect could be a mixture of counting traffic **flows** and **qualitative data**, such as taking photos and using questionnaires. However, make sure you know which method will be the focus of the assessment in the year you take the examination and use the appropriate chapter. See Figure 3.

Section 3 describes six geographical concepts and how they might be investigated through fieldwork. Make sure you know which concept will be the focus of the assessment in the year you take the examination and use the appropriate chapter. See Figure 3.

Figure 3 Which method and which concept will be assessed?

Year of the exam	Method	Concept
2018	Use of transects	Sphere of Influence
2019	Measuring flows	Mitigating risk
2020	Qualitative data	Sustainability
2021	Change over time	Cycles and flows
2022	Qualitative data	Place

How is fieldwork assessed?

An understanding of fieldwork is assessed in an examination at the end of your GCSE. Your fieldwork will have a focus that is set by Eduqas, your exam board. The focus for the assessment changes each year and is shown in Figure 3. You will sit an examination paper. It is called Component 3 and is worth 72 marks.

- Parts A and B of the examination will assess your understanding of the enquiry process and your own fieldwork. There are 36 marks altogether.
- Part C of the examination will assess your ability to make a decision about the same geographical concept that you have been studying in your fieldwork. This part also has 36 marks.

Figure 4 What does Component 3 look like?

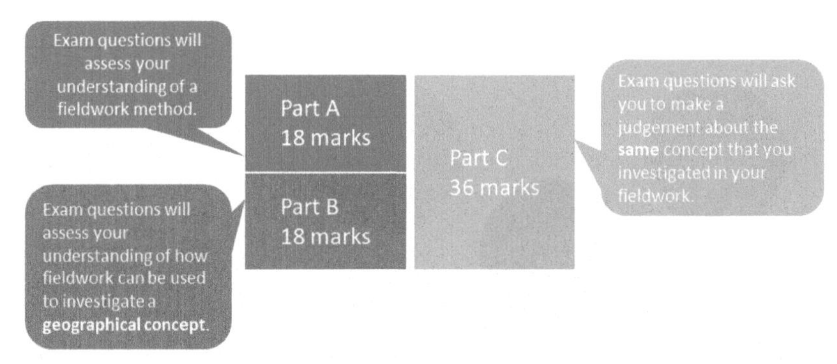

How can you prepare for the examination?

As a GCSE student you must do two fieldwork enquiries. You should keep notes about each of your fieldwork enquiries in a notebook or portfolio. You should:

- organise your portfolio using six headings: one for each stage of the enquiry process;
- discuss the main **methodologies** you used to collect data. What were their strengths and weaknesses?
- make sure you justify your decisions. For example, can you explain why you used a particular sampling strategy or presentation technique? What made it suitable? Keep notes about the decisions and choices you made;
- use different techniques to process and present data. Discuss which ones work best;
- think about the concept you investigated. Make sure you understand it and how it could be investigated in other ways through fieldwork.

Revision tips

The exam will have two types of questions.

- Some questions will be about your own fieldwork.
- Most questions will provide you with some information about how another group of students did their fieldwork. You can apply your experience of your own fieldwork to answer these questions. To help prepare for these, you can practise the questions at the end of the relevant chapters in Section 2 and Section 3. These are in a similar style to the questions you could find in Parts A and B of the Component 3 examination.

Questions in the exam will test your ability to evaluate and make decisions about fieldwork. The questions could be about any of the six stages of the enquiry process. The examiner will want to see evidence that you can think for yourself. For example, can you explain why some ways of presenting your data might be more effective than other ways? Can you evaluate the strengths and weaknesses of your sampling techniques? Figure 5 reminds you of key words to use when answering questions about your own experience of fieldwork in Parts A and B of Component 3.

Figure 5 Key words to use when answering questions about your own experience of fieldwork

You can evaluate your aim using the mnemonic SMART (Specific, Measurable, Achievable, Realistic, Time-based). Make a copy of the bipolar survey shown in Figure 6. Use the bipolar statements to evaluate your own enquiry. Once you have completed the survey, you could:

- explain why one aspect of your aim was successful;
- explain why another aspect of your aim had limitations;
- suggest and justify how your aim could have been improved.

		5	4	3	2	1	
S	The enquiry had a clear aim and specific focus. I know what I was trying to prove.						The aim was vague. It lacked focus. I didn't know what I was trying to prove or disprove.
M	I could measure all outcomes of the fieldwork. I used a range of techniques to collect qualitative and quantitative data.						I struggled to measure any outcomes of the fieldwork. I wasn't sure which techniques to use.
A	The enquiry was achievable and safe. Primary and secondary data was available to provide the evidence I needed to meet my aims.						There wasn't enough of the right kind of data available to me so I couldn't find the evidence I needed to prove / disprove my aims.
R	The aims were realistic. I had the right data collection skills and accurate equipment.						The aims were unrealistic because I didn't have the right skills or equipment to collect the data I needed.
T	I had enough time to collect data across a range of places / time so that I could see patterns and trends.						I would have needed more time to collect enough data to see any patterns or trends in the data.

Figure 6 How to evaluate your aim

Fish bone diagrams are a useful way to analyse the factors involved in any process (see Figure 7). Drawing one should help you to evaluate each factor that may have affected your data collection. The first thing to do is to list all of the factors that might have had any effect on how data was collected – these become the bare bones of your diagram. Then, you can think about the strengths and weaknesses of each of these factors. It may help you to identify some simple ways that your data collection could have been improved.

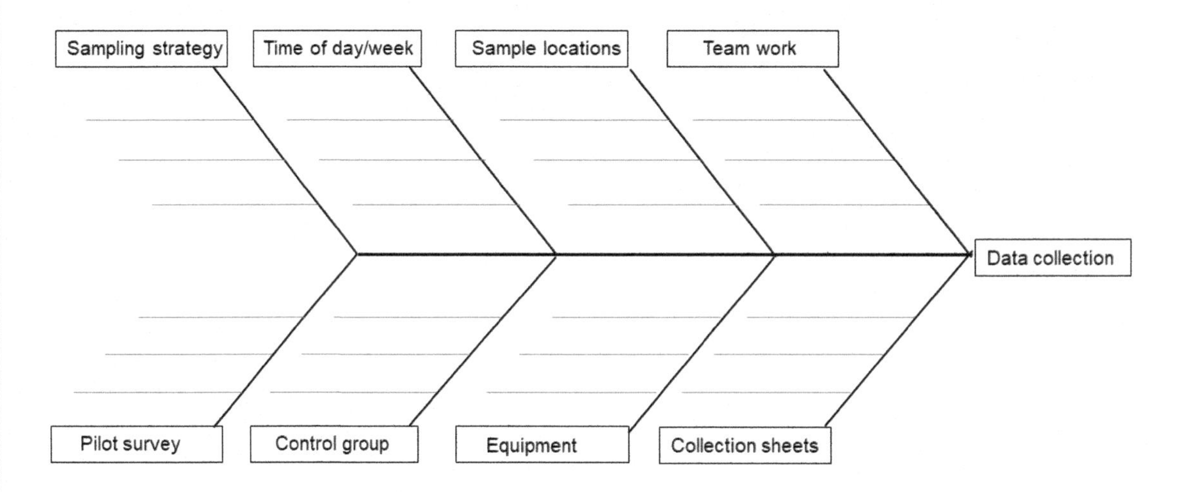

Figure 7 Use a fish bone diagram to think about the strengths and weaknesses of each factor that contributed to the process of collecting the data

Summary

- Find out which Method and which Concept will be assessed in your Component 3 exam.
- Organise your fieldwork portfolio using the six stages of the enquiry process.
- Evaluate your aim. How SMART was it?
- Think about the main sampling methods you used, why you used them, and whether or not they were successful. Use the fish bone diagram to help you.
- Make notes about which data processing techniques worked well and why you chose them.
- Think about how successful your fieldwork was in identifying trends, patterns or connections.

Chapter 1

Creating an aim for fieldwork

SECTION 01

Making decisions about your enquiry

Learning objectives

In this chapter we will explore:

- how to identify an overall aim for your fieldwork;
- creating a hypothesis or enquiry question to break this aim down into manageable chunks.

First you need to have an **aim** for your fieldwork. You can't just start your fieldwork by going outside and measuring things or counting stuff. Fieldwork can involve a lot of counting and measuring but the counting has to have a purpose.

In order to create an aim, you need to think carefully about the concept you are investigating. You also need to know where your fieldwork is going to take place. Will you be working in a river environment, a beach, or in a town? Thinking about the location and concept will help you create an aim for your fieldwork.

Thinking of useful geographical questions is an important part of any fieldwork enquiry. You can see how knowing about the concept and location could lead to an aim for your fieldwork in Figure 1.

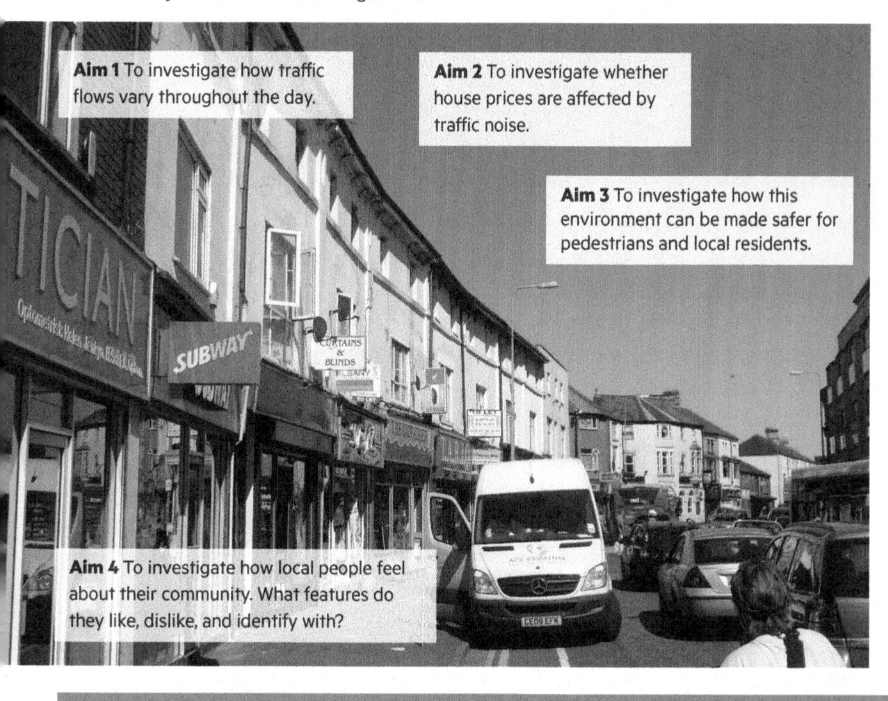

Figure 1 Four possible aims that could be developed in this location. Cowbridge Road East, Cardiff

Activities

1 **Study Figure 1.**
   a) If you were able to visit this location on a fieldtrip:
      i) how might counting the traffic help you answer one or more of these aims?
      ii) what other data could you collect to help you answer each of these aims?
   b) Each of the aims is connected to one of the concepts listed in Figure 1 on page 3. Match the aims to the concepts.

What makes a good enquiry question?

You can use questions to break down your aim into manageable chunks but what makes a good enquiry question? To help you ask your own enquiry questions you could use the question starters that have been added to Figure 2.

Figure 2 The 7Ws and an H. Happisburgh, Norfolk

Figure 3 gives three different styles of question.

- Questions on the left are useful but not very demanding because the answers are probably very straightforward.
- The questions in the middle are better. They are more useful and interesting as enquiry questions because they are about the relationship between two variables.
- The questions on the right are more interesting again and could provide excellent enquiry questions because you may find a variety of potential answers. However, this type of question needs to be phrased very carefully and you may need to gather a great deal of evidence to answer them fully.

Figure 3 Styles of enquiry question

	→ increasing complexity of question →		
	Questions that have a straightforward answer	Questions about connections	Questions that have a complex answer
Examples of question starter	When? Where? How many? Which?	How does? What effect would?	Who benefits? What ought? Who should?
Examples of questions	Which is the busiest shopping street? When is the busiest time of day? How many cars pass in 5 minutes?	How does wind speed vary with altitude? What effect would traffic noise have on house prices?	Who would benefit most if flood protection was improved? What ought to be done to make this community more sustainable?

Choosing a hypothesis

Just like a science experiment, you can sometimes use geography fieldwork to prove or disprove a **hypothesis**. A hypothesis is a provisional idea which can be proven to be correct or incorrect based on the factual evidence collected in your fieldwork. An example of a hypothesis that could be investigated in Figure 1 is:

> ***During the morning rush hour, more traffic is travelling into the city centre than out of it.***

Once you have chosen a suitable hypothesis for your fieldwork you should state the null hypothesis.

This is what you would discover if the hypothesis was incorrect. For Figure 1, the null hypothesis would be:

> ***'There is no difference in the amount of traffic travelling in and out of the city during the morning rush hour'.***

Setting a SMART aim

Is your aim a good one? Is it suitable? An aim shouldn't be too simple. For example, 'How many cars travel down this road in 5 minutes?' is not an aim. This simple question is part of the broader aim of investigating patterns of commuter movement. However, an aim shouldn't be too complex either – otherwise you may not be able to achieve it. For example, 'Cowbridge Road East is a perfect example of a modern sustainable residential community' would be very difficult to prove or disprove because the word 'perfect' is so difficult to measure. Your aim should be SMART.

Is your aim SMART?

- **Specific** – does the aim have a clear focus?
- **Measurable** – can you collect and count data?
- **Achievable** – will it be safe? Will you need a team of people? Is data available?
- **Realistic** – have you got the right skills and/or equipment?
- **Timely** – will you have enough time during the fieldtrip to collect all of the data?

Take a virtual visit

Another good way to prepare for your fieldwork is to make a virtual visit. Use the internet to see the environment you will be working in before the day of the fieldtrip. This has a number of advantages. It will help you to:

- pose enquiry questions or set a hypothesis that relates to the actual place that you will visit:
- plan where you could collect the data:
- consider potential risks of the site and plan your visit so you stay safe.

Activities

1 Study Figure 2.

a) Write 10 enquiry questions using as many different question starters as you can.

b) Evaluate your questions using Figure 3. Which are your best 3 questions? Are they SMART? Are they complex?

Chapter 2

How is evidence collected?

Learning objectives

- Understanding the difference between qualitative and quantitative data.
- Understanding the difference between primary and secondary data.

Types of data

In a fieldwork enquiry we need to collect data that will provide the evidence we need to answer our aims. We describe data as either **quantitative** or **qualitative**.

Quantitative data is something you can count, or measure. Some evidence that you can measure will need specialist equipment. For example:

- wind speed with an anemometer;
- river velocity with a flow meter;
- slope angles with a clinometer.

However, lots of quantitative data can be simply counted such as the number of vehicles on a road, the number (and type) of shops, and the number of pedestrians in a shopping centre.

Qualitative data is something that helps us to describe a place or geographical issue with words or pictures rather than by measuring with numbers. It can include interviews, making field sketches, or taking photographs, videos, or audio recordings. Section 2 Chapter 9 describes some specific techniques you can use.

Figure 1 Examples of quantitative and qualitative data

Quantitative data	Qualitative data
Depth and width of a river Wind speed House prices The size of pebbles on a beach	Whether pedestrians feel safe in the city centre Whether people prefer shopping in the high street or out of town What different groups of people think about the risk of flooding

Activities

1 **Make a copy of Figure 1. Decide whether each statement below is quantitative or qualitative and then add it to the correct column.**
The number of wave crests that break each minute on a beach.
Photos of features of an urban environment that provide evidence of sustainability.
The reasons people move to live somewhere new.
A field sketch of a coastal landform.
The angle of slope of a beach.
Photos of features of a shopping centre that are liked/disliked by shoppers.
Noise (in decibels) of traffic from a road.
An audio recording of a busy shopping street.
Infiltration rates of water soaking into different soils.
Wind speeds at different locations up a hill.

What is the difference between discrete and continuous data?

Discrete data and continuous data are two different types of quantitative data. **Discrete data** can be counted. We could count how many people are in a shopping street, or how many vehicles pass along a street. That would be discrete data. By contrast, **continuous data** is measured. We can measure how quickly the river is flowing, or how much noise the traffic is making. That would be continuous data. Distance, velocity and time are all examples of continuous data that are often measured in geography fieldwork.

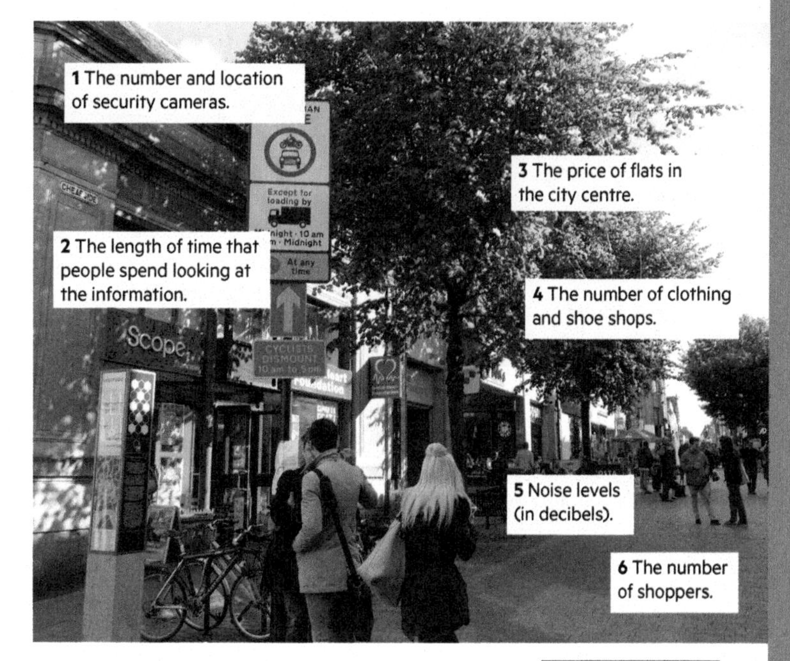

Figure 2 Examples of discrete and continuous data that could be collected in a retail environment, Lancaster

What is the difference between primary and secondary data?

The data you collect during the fieldtrip, such as the number of vehicles passing, or the answers to a questionnaire, is known as **primary data**. Sometimes you will be able to answer your aim by only collecting primary data. However, a lot of data has already been collected and is available in books and on websites. This is **secondary data**. There are lots of forms of secondary data that might be useful to help you answer your fieldwork enquiry. With careful research, you can find:

- photographs that show your fieldwork location, both now and in the past - you could annotate these to show how the location has changed;
- blogs – you can analyse people's viewpoints;
- maps - for example, the Environment Agency Flood Map would be useful if your fieldwork is located in a town that is at risk of river or coastal flooding;
- tables of data – which you can use to draw your own graphs.

Strengths and limitations of primary and secondary data		
	Strengths	**Some limitations**
Primary data	You know exactly how the data was collected.	Your sampling strategy may not be representative (see page 13-14).
Secondary data	It's much quicker to download secondary data than collect your own primary data.	You may not know how or when it was collected and the figures may be out of date.

Activities

2 Study Figure 2.

a) Sort the quantitative data into discrete and continuous types.

b) Suggest three examples of qualitative data that could be collected here.

Learning objectives

- Understanding the difference between random, systematic, and stratified sampling.
- Knowing how to choose the right sampling strategy.

Why do I need a sampling strategy?

Imagine you want to see how the amount of traffic varies throughout the day in a small commuter town. It wouldn't be possible to count every single vehicle travelling along every road for 24 hours. It would take too much time. Also, you would need your friends to agree to help you because you can't watch every road at the same time. So, instead of trying to collect every bit of data, you will need to collect a **sample** of data. If collected properly, your sample will accurately represent the patterns or trends shown in all of the data.

What sampling strategies are available?

There are various ways you can select your sample. Figure 3 summarises four main strategies that you could use.

Figure 3 Four commonly used sampling strategies

Sampling strategy	How it works	Examples – you collect data...
Systematic	Your sample is chosen at regular intervals. Every n^{th} item is sampled.	• at points that are equally spaced (for example, every 10 metres) along a line (transect) • every hour • by interviewing the 10th, 20th and 30th person who passes you
Random	Your data is sampled at random intervals. Every item has an equal chance of being sampled.	• at places that have been chosen by using tables of random numbers to generate grid references • in streets chosen by putting all street names into a bag and pulling out a sample of names at random
Stratified	Your data falls into two or more categories. You make sure that your sample reflects the proportions that fall in each category.	• by sampling plants and wind speeds in a sand dune. You choose 50% of your sampling points on slopes that face the sea and 50% of your sample points on slopes that face away from the sea • you are sampling how students at your school feel about a geographical issue. 45% of the students are in Years 7 to 9, 40% are in Years 10 and 11, 15% are in the sixth form. You make sure that your questionnaire is given out in these proportions.
Opportunistic (also known as convenience sampling)	Your data is scarce so you take the opportunity to collect data whenever or wherever you can.	• by asking a questionnaire of every tourist you can find • you intended to systematically sample downstream along the course of a river. However, a lot of the sites you wanted to sample were on private land so you sampled wherever there was a public footpath next to the river.

Sampling along a line

Figure 4 By collecting data at regular intervals across this river channel the students are sampling systematically along a line

In many fieldwork enquiries a hypothesis will try to investigate how something changes over distance. For example:

- rivers get wider as you go downstream;
- noise reduces as you get further away from a busy road;
- house prices get higher as you get closer to the park.

Each of these hypotheses could be investigated by sampling data along a line. You can use any of the four main sampling strategies while you work along a line. Sampling along a line is often described as a transect. You can read more about transects on pages 64-69.

Sampling across an area

People say that if you can map it, it's geography! This is because geographers are interested in how features are distributed. The patterns we see on a map are known as **spatial** patterns. If you sample along a line you are sampling in one dimension. You will see how a pattern changes with distance but you will not be able to see a spatial pattern. To see a spatial pattern (and then draw a map) you will need to sample in two dimensions. You could do this by:

- sampling along several lines, for example, four lines radiating out from the town centre towards the suburbs;
- using a map of a town to identify sampling locations at each grid intersection. This would be a form of systematic sampling.

Which is the best sampling strategy for me?

Each sampling method has advantages and disadvantages which makes it more suitable in some situations than others.

- Random sampling provides an unbiased sample. If the full size of the thing that is being sampled is quite small (like shops in a town, or students in a year group) this method should create a sample that represents the whole accurately – we say that it is **representative** of the whole.
- Systematic sampling is much quicker and simpler than random sampling. It's a better method to use if the thing that is being sampled is quite large (like a questionnaire of shoppers, or plants along a line through an ecosystem). Systematic sampling across an area should generate enough data to draw a map.
- When the thing that is being sampled is very large (for example, pebbles on a beach), you can combine random and systematic sampling strategies (see pages 14-15).
- Opportunistic sampling is really a method that you choose because it is not practical to choose another method. It may be quicker or easier than other methods but the results may not be representative.

http://www.textfixer.com/numbers/random-number-generator.php One of many online tools that will generate random numbers for you.

Learning objectives

- How to sample data using a quadrat.

Sampling data using a quadrat

A quadrat is a simple square frame that you place on the ground. It gives you a shape that you can sample inside. The frame is usually quite small – often 50cm by 50cm (which is a quarter of a square metre) - like the one shown in Figure 5. However, you can also use tent pegs and string to mark out a larger frame – say 2 metres by 2 metres. The frame is often divided into smaller squares with wire or string to make it easier to identify sample locations within the square.

Quadrats are used to help sample a very large population by breaking down a large space into smaller, more manageable spaces – concentrating on the space within the frame of each quadrat. They are often used to sample plants (for example, in a sand dune ecosystem) or pebbles (on a beach).

There are two main ways to use a quadrat.

- **One** Sample objects within the quadrat using the points where the internal wires/strings cross – in other words, at the intersections. Use them systematically, for example, by sampling pebbles at every intersection, as in Figure 5.
- **Two** When you are sampling types of vegetation, it is usual to estimate the area covered by each type of plant within the quadrat, as in Figure 6. Do this by counting the squares (the spaces between the wires/strings). The quadrat in Figure 6 is divided into 100 squares so it is easy to estimate the percentage distribution of each plant as each square represents 1%. This style of quadrat works best with low growing plants. The quadrat in Figure 5 has five rows and five columns of spaces between the strings. That makes 25 small squares inside the quadrat. The quadrat represents 100% of the sample area, so each small square represents 4% of the whole (because 100 divided by 25=4). This style of quadrat could be used to estimate percentage area covered by taller plants.

Quadrats are often used to sample vegetation along a belt transect. For more information on this technique, you can use page 64.

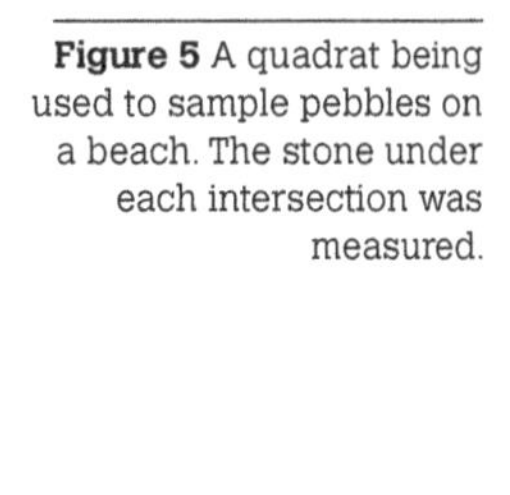

Figure 5 A quadrat being used to sample pebbles on a beach. The stone under each intersection was measured.

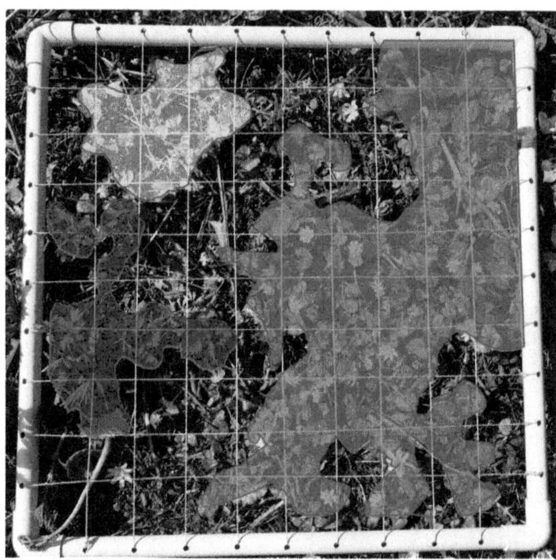

Figure 6 Estimating areas within the frame of the quadrat. The percentage of the quadrat occupied by each species of plant was estimated. Three different species of plant can be identified in this quadrat. The second image has been tinted to show the percentage area covered by each plant more clearly.

Control groups

Quadrats are often used to measure how vegetation changes in relation to another variable, for example, how vegetation varies:

- across a sand dune with increasing distance from the sea;
- across a footpath with variable amounts of trampling.

When you are looking for variations like this it is a good idea to see what the vegetation is like by sampling in an area unaffected by the variable. We call this a **control group**. It allows you to compare the results of the area that you are investigating (for example, the plants that have been trampled on the footpath) with the normal vegetation of the area.

Dos and don'ts of using quadrats

Do:

- ✓ Select locations for the quadrat using a systematic, random, or stratified technique.
- ✓ Take care to avoid damaging the vegetation where you are working.

Don't:

- ✗ Throw the quadrat over your shoulder. This doesn't make the sampling random and it could be dangerous!
- ✗ Select pebbles from within the quadrat at random by just picking some out. Your eye will be drawn to the larger, or more interesting, looking pebbles. This means your sample won't be representative.

Strengths and limitations of using quadrats	
Some strengths	**Some limitations**
Quadrats are useful for sampling a very large population (like pebbles or plants) that is spread over a large area. You can use quadrats to record the presence or absence of a variable (whether it is in the quadrat or not) and also the frequency of a variable (how much of the variable is present). Quadrats are relatively cheap pieces of equipment and simple to use.	It can be difficult to be accurate about the percentage of vegetation in a quadrat if the plants are different heights. Taller plants grow over shorter ones so it is possible to underestimate the percentage of the quadrat filled by smaller plants.

Learning objectives

- How to use a clinometer.

Using fieldwork equipment

A lot of data can be collected without the need for any special equipment. You will not need special equipment to collect qualitative data through **bipolar surveys**, **Likert surveys**, and questionnaires, although you will need to design a data collection sheet before the fieldwork begins (see pages 18-19).

Quantitative data that is continuous will need to be measured and some will need specialist fieldwork equipment. Wind speed, for example, needs to be measured using an anemometer, like the one in Figure 7. However, other measurements are easier to make. For example, in the river fieldwork seen in Figure 4 on page 12:

- the width of the river is measured using a measuring tape;
- the depth of water is measured using a metre rule;
- the velocity of river flow can be calculated by floating a dog biscuit downstream for 10 metres and using the stop watch app on your smart phone.

Figure 7 An anemometer is used to measure wind speed

Using a clinometer

A clinometer is used to measure angles from the horizontal so, in fieldwork, it can be used to measure the angle of a slope. A traditional clinometer includes a swinging arm or weight. As the clinometer is tipped up to the same angle as the slope, the weight points vertically downwards so that you can measure the angle. Alternatively, you could download a free clinometer app onto your smartphone, like the one shown in Figure 8.

To measure the slope of a beach, like the one in Figure 8, you will need:

- a clinometer (or an app on your smartphone);
- two people;
- a long measuring tape;
- two ranging poles (although these are not essential);
- a data recording sheet.

Step One Set up a transect, up and down the beach, using the long measuring tape.

Step Two Identify the **breaks in slope** along the transect. These are places where the gradient of the slope suddenly changes. The photograph in Figure 8 has been coloured so that you can see the breaks in slope.

Step Three Start measuring at the bottom of the slope. If you do not have ranging poles, work with someone who is the same height as you. Get your partner to stand at the next break in slope. Hold the clinometer up to your eye and look along it into your partner's eyes. The angle recorded by the clinometer is the angle of the slope.

Step Four Record the angle (in degrees) and the distance between the start of the transect and the first break in slope (in metres) on your data recording sheet.

Step Five Later you can present your data as a line graph using a ruler and protractor. This graph will represent the shape of the **beach profile.**

Step Six Move up the transect to the first break in slope. Your partner should also move up the transect and stand at the next break in slope. Take the measurements and repeat this process until you have completed the transect.

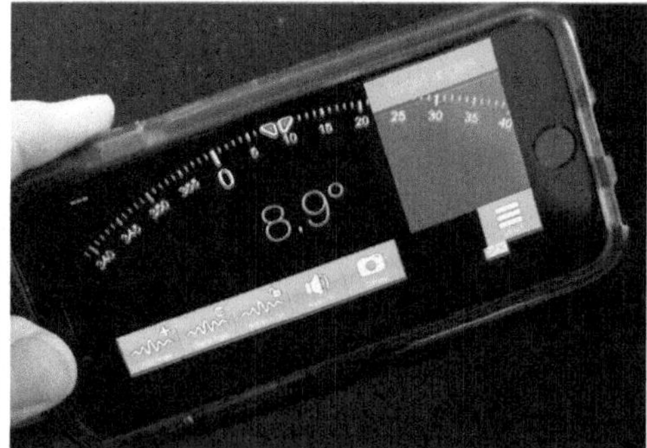

Figure 8 Using a clinometer to measure slope angles on a beach. The dotted black line represents the transect. Notice that the red dashed line is parallel to the slope.

Dos and don'ts of using a clinometer

Do:

- ✓ Check that the tide is going out before you start. Then you can start the transect at the lowest part of the beach without any danger.
- ✓ Take three angle readings at each break in slope and take a mean. Clinometers that use a swinging arm can be affected by the wind. Taking an average will improve the reliability of your results.

Don't:

- ✗ Work with someone who is a different height unless you are using ranging poles. If you line up the clinometer with a colleague's eye and they are much taller than you are then your angle reading will be too large.

Figure 9 The data below was collected by students measuring slope angles on a beach in the UK

Site	Distance (metres) between sites	Angle (°) between sites
1 - 2	10	5
2 - 3	22	8
3 - 4	20	10
4 - 5	26	15
5 - 6	22	8
6 - 7	30	6
7 - 8	40	10
8 - 9	50	15
9 - 10	20	20
10 - 11	40	5
11 - 12	30	3

Activities

1. **Present the information from Figure 9 as a line graph to show the beach profile. You will need graph paper and a protractor to do this accurately. Plot distance from the sea on the horizontal axis. For sites 1-2 draw a straight line of 10 metres length from the origin (0) at an angle of 5°. From the end of this line draw a line of 22 metres length at an angle of 8°. Continue until all the data is plotted.**
2. **Annotate your line graph to show significant features of the beach profile, such as breaks in slope.**

Learning objectives

- Knowing when to use a pilot survey.
- Understanding how to design data collection sheets.

Designing data collection sheets

Now you know what data you want to collect there is one more important job to do before you go on your fieldtrip. You will need a quick and methodical way to record the evidence that you collect on your fieldtrip – so you need to design data collection sheets.

Your data collection sheet needs to be easy to use. Always record when and where the data was collected at the top of the sheet. Data can be broken down into categories (for example, size of pebbles, or age of people) and you can then add boxes or tables to your sheet so you can record results in each category. Using a tally mark, like Figure 10, is a good way to collect information that you are counting.

Traffic survey

Location HIGH STREET

Date 19/05 Time 11 am

Type	Tally	Total
Cycles	||||	4
Motorbikes	||	2
Cars	~~||||~~ ~~||||~~ ~~||||~~ ||||	19
Vans	~~||||~~ ~~||||~~	10
Lorries	~~||||~~ |	6
Buses	~~||||~~	5

Figure 10 Use tally marks when you are counting things

Continuous data, such as wind speed and noise levels, can vary considerably as you are measuring them. It's a good idea, therefore, to create several boxes for these values on your data collection sheet, like in Figure 12. Take 5 readings, one every 30 seconds. Then take an average.

Lots of students start a questionnaire, or bipolar survey, by recording the gender and age of the person they are talking to but this often isn't necessary. You should only do this if age or gender are factors that you want to analyse as part of your enquiry. For example, it would be essential to record the age of the respondents in your questionnaire if your enquiry question was *'Are older people more aware of risk in the environment than younger people?'*.

Beach survey data collection sheet

Date ......

Site 1

Pebble sizes (mm)	Number of pebbles
Less than 10	
10-20	
25-40	
40-50	
More than 50	

Figure 11 There are three errors in this data collection sheet. Can you spot them?

Location of noise readings ....

Time they were taken ....

Readings (one taken every 30 seconds)	Noise level (dB)
1	
2	
3	
4	
5	
Total	
Average (total/5)	

Figure 12 Design your data collection sheet so that several readings can be taken because noise readings may vary a lot

Dos and don'ts of data collection sheets

Do:

- ✓ Create a space at the top of the data collection sheet to record the time, date, and location of your survey
- ✓ Use boxes, tables, and tally marks to collect data methodically and quickly
- ✓ Each column in your data collection table must have a heading. Make sure you also record the units of measurement in the heading, for example, Depth of water (cm).

Don't:

- ✗ Record the age and gender of the people in your questionnaire unless you are going to use this information in your analysis
- ✗ Create categories that overlap each other or which miss out some individuals

Pilot surveys

If you are going to use a questionnaire (see pages 76-77) you will want to use some closed and some open questions. Open questions allow the person you are interviewing to say whatever they want in response. The following are examples of open questions that you might use in a fieldwork enquiry about place and identity (see pages 92-93):

1. Which features do you like best about this town?
2. What makes this town different from others you know well?
3. What gives this place its unique identity?

Using open questions like these have advantages but it is time consuming to record all the possible responses that people might give. So, it's a good idea to use a **pilot survey** with a few people to see what kinds of answers you are likely to get when you carry out the full survey. After the pilot survey, you can look at the answers that people gave and put them into categories. So, in a pilot survey, the open question *'Which features do you like best about this town?'* was asked. Imagine that:

- 5 people said friendly people;
- 4 people talked about history or historic buildings;
- 3 people said that good schools were an important feature of the town;
- 3 people said they liked the multicultural atmosphere;
- 1 person talked about locally owned (independent) shops.

As a result of this pilot survey you can now design a data collection sheet for your full questionnaire that looks like Figure 13.

Figure 13 Using a pilot survey to create option boxes for the data collection sheet

1. Which features do you like best about this town?
Tick up to 3 boxes.

The people	☐
The town's history	☐
Buildings/architecture	☐
Local schools	☐
Cultural activities	☐
Multicultural atmosphere	☐
Independent shops	☐
Others	☐

If others, please describe the feature(s) ...

Activities

1. **Study Figure 11. Identify three errors in the design of this data collection sheet.**
2. **Explain why it would be important to use a pilot survey to create the data collection sheet in Figure 13 rather than just making up the categories.**

Designing data collection sheets for an Environmental Quality Index (EQI)

Urban fieldwork often involves the collection of environmental evidence. For example, you might want to collect data on:

- safety of the environment (quality of pavements, traffic calming, CCTV);
- pollution (noise, litter, graffiti);
- access (safe places to cross roads, cycle paths, ramps into shops);
- green spaces (amount of vegetation, how well they are maintained and used).

In order to collect this kind of data you will need to design an **environmental quality index (EQI)** sheet. Precise criteria should be used in your sheet. These are statements that are used to quantify each feature of the environment being assessed. This will enable you to get consistent results if several people are collecting the data. Each criterion can then be numbered using, for example, a scale of 1 to 5. This means you can process the data later by calculating means and drawing graphs and maps. An example of this is shown in Figure 14.

It is a good idea to carry out pilot surveys to assess the effectiveness of your EQI sheet before you use it on the fieldtrip. This could be done near to the school, or even in the school grounds, so that the criteria can be discussed with your colleagues. You can then amend the EQI data collection sheet if necessary before the actual fieldtrip.

Don't forget to add the following to your data collection sheet:

- a space to record the location of the survey;
- a space to record the time and date of the survey.

Figure 14 Use precise criteria to assess each feature in an EQI survey

Feature	Criteria				
Vegetation	5	4	3	2	1
	Plenty of trees. 1+ tree per 20m.	Some trees. 1 tree per 20-40m	Few trees. 1 tree per 40-80m	Sporadic trees. 1 tree per 80-100m	No trees or greenery.
Litter	5	4	3	2	1
	Occasional litter. One item every 50m+.	Hardly any litter. One item every 11-50m.	Some litter. One item every 6-10m.	Lots of litter. One item every 1-5m.	Abundant litter. More than one item per metre.

Using your phone to measure noise levels

You can use your smart phone to measure noise. You will need to download an App that measures decibels (dB) – there are several free ones available.

Normal conversation is about 65dB. The decibel scale is logarithmic. This means that 75dB is 10 times louder than 65dB and 85dB is 100 times louder than 65dB. Noise on a busy city street will be between 75dB and 85dB.

Adding weightings to an EQI

It is possible that the features included in the survey may not have equal importance. For example, local residents may tell you that they are much more affected by traffic noise and litter than they are by graffiti. If so, you can increase the weighting of the features that are considered to be more important. Do this by multiplying them by a factor, as in Figure 15.

Figure 15 Use weightings to reflect the view that some features of the environment are more important than others

Feature	Criteria				
	5	4	3	2	1
Vegetation	Plenty of trees. 1+ tree per 20m.	Some trees. 1 tree per 20-40m	Few trees. 1 tree per 40-80m	Sporadic trees. 1 tree per 80-100m	No trees or greenery.
	5x2=10	4x2=8	3x2=6	2x2=4	1x2=2
Litter	Occasional litter. One item every 50m+	Hardly any litter. One item every 11-50m	Some litter. One item every 6-10m	Lots of litter. One item every 1-5m.	Abundant litter. More than one item per metre.
	5x3=15	4x3=12	3x3=9	2x3=6	1x3=3
Traffic	Occasional traffic. Less than 5 cars per minute.	Hardly any traffic. 5-9 vehicles per minute.	Some traffic. 10-15 vehicles per minute. All cars	Lots of traffic. 16-20 vehicles per minute. Lorries and cars.	Traffic is a nuisance. More than 20 vehicles per minute. Lorries and cars.

Strengths and limitations of EQI surveys?	
Some strengths	**Some limitations**
EQIs can include a wide range of environmental factors. They provide quantitative data which can easily be presented. Scores collected by different groups of students can be averaged to improve reliability. Precise criteria for each feature will increase reliability and comparability of results. If EQIs are collected along a transect, trends over distance can be analysed. If EQIs are collected at different places (in two dimensions) the numbers can be mapped and spatial patterns can be analysed.	Scores given may be affected by the time of day, week, or year. Scores for each feature will be inconsistent if groups of students who are using the EQI haven't agreed the criteria carefully first.

Activities

1 Study Figure 14.

a) Explain why the scoring gives 5 marks for 'plenty of trees' and the same score for 'occasional litter'.

b) Write a set of criteria for:

i) Graffiti/vandalism

ii) Condition of pavements

iii) Access to shops/services

c) Explain why graffiti/vandalism, condition of pavements and access to shops might be considered to be useful indicators of the quality of an urban environment.

2 Study Figure 15.

a) Describe how you might use a pilot survey to decide how to set these weightings

b) Explain why weightings like these are useful.

Collecting qualitative data - making images during fieldwork

Learning objectives

- Knowing when it would be useful to photograph or sketch.

You can collect images as evidence of your fieldwork by:

- taking photographs;
- making field sketches. These are simple outline drawings of the landscape;
- creating sketch maps. These are simple maps of the area in which you do your fieldwork. They do not need to be to scale.

Each of these images can be labelled or annotated later when you are analysing your evidence (see page 38).

You might also take video during your fieldwork to collect evidence of, for example:

- the complex way that people move through pedestrianised streets;
- the ebb and flow of traffic during rush hour.

The evidence shown in photographs, videos, or sketches is a form of **qualitative data**.

When should images be taken?

If you decide to make images during your fieldwork it is important that you make sure that each image serves a purpose. For example, you might:

- use an image to provide an overview of the main features of the study area. Field sketches and sketch maps are particularly useful for this because you can concentrate on just showing the main features of the landscape that are relevant to your fieldwork;
- take a photograph to illustrate the way that data has been sampled. You may be able to annotate this image to analyse the strengths and limitations of the data collection technique;
- take a series of photographs to provide evidence of individual features that are relevant to your fieldwork. For example, you might want to take photographs of the features that people have identified as being important in a questionnaire or longer interview.

http://www.gridreferencefinder.com/ Use this simple website to find six figure grid references for your fieldwork locations and photos.

Dos and don'ts of taking photos

Do:	Don't:
 Provide a reference for the location of any photo that you take for your investigation. An OS grid reference is often best. See the web link above.	 Take photographs unless you can analyse them or use them somehow. You are not going to make a scrapbook.

Why draw a sketch map of your fieldwork area?

It might be easy to download a map of your fieldwork area from a website. However, these maps are often cluttered and may be difficult to use. You might decide to use a sketch map instead because you can:

- leave out unnecessary details;
- just include the main features;
- label the features that are important to your fieldwork.

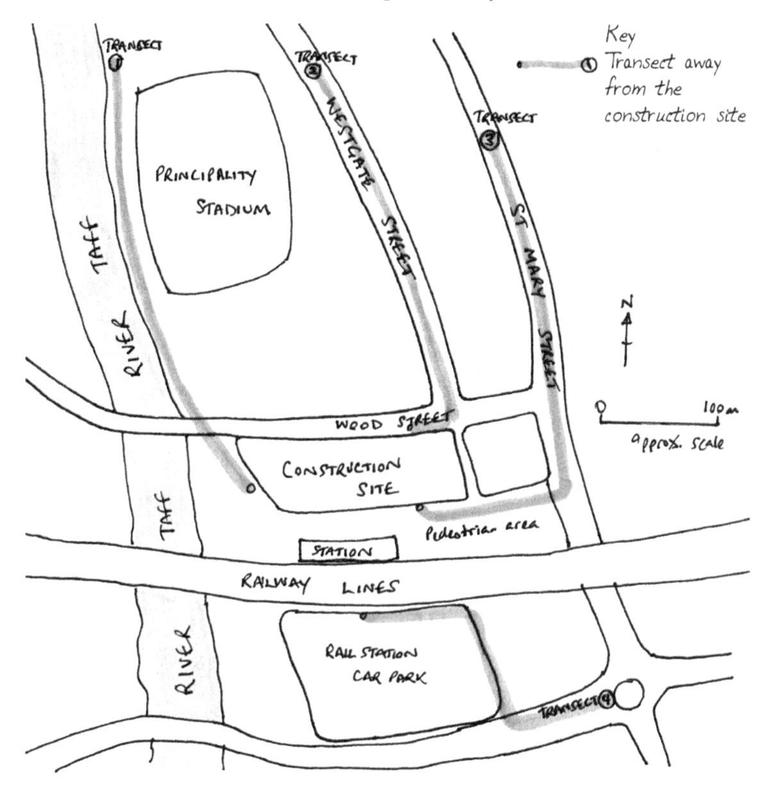

Figure 16 A sketch map of a fieldwork location can focus on just the essential features. This example shows a potential fieldwork location in Central Cardiff

Figure 17 Apps on your smart phone can be used to record date, location, and altitude. This is the building site shown in the centre of Figure 16. An altitude app on the smart phone used the phone's camera and recorded altitude and location in the bottom left of the photo

Dos and don'ts of making sketch maps

Do:	Don't:
✓ Give your map a heading, a north arrow and an approximate scale line; ✓ Make a key to match any colours or symbols on your map.	✗ Add too many details and clutter the map.

Activities

1. **Figure 16 was created at the postcode CF10 1EP.**
   a) Use a search engine to find a map of this postcode and its surroundings.
   b) Compare the search engine map to the sketch in Figure 16. What features have been left out of the sketch? Suggest why this make the sketch more useful.
2. **Suggest why it is important to record time, location and altitude as evidence when making a fieldwork image.**

Learning objectives

- Being aware of secondary sources that can be useful in the field.

Using secondary sources to help collect data

A lot of data is available in books and on websites. This is **secondary data** and it can be used at several stages of the enquiry process as shown in Figure 18.

Figure 18 How secondary data may be used in fieldwork

Stage of the enquiry process	How to use secondary data
Planning	Photographs of your fieldwork location from the internet can be used at the planning stage to help you create a hypothesis or pose an enquiry question.
Data collection: Sampling procedures	Tables of data can be used to help you select a sampling strategy. An example is shown on page 27.
Data collection: Collecting qualitative data	Blogs can be used to collect evidence of people's viewpoints.
Data collection: Collecting quantitative data	Many websites contain huge quantities of data. The use of these **big data** sets is described on pages 26-27.
Data presentation	Tables of data from websites can be used to draw maps and graphs.
Analysis	Websites may include graphs and maps that are relevant to the aims of your fieldwork. Select some and annotate them to help your analysis.

Dos and don'ts of using secondary data

Do:

- ✓ Select secondary data carefully. Only present and analyse data that will be useful to answer your aim.
- ✓ Acknowledge where your evidence has come from. Do this by giving the URL of the webpage where you found the data. Then you will be able to find the source again if you need it.
- ✓ Consider the reliability of the source. Can the evidence be trusted?

Don't:

- ✗ Copy photographs or other evidence from the internet into your portfolio if you don't intend to process or analyse it. You can collect too much stuff which will make your portfolio difficult to work with.

Using your smart phone during fieldwork

Some apps, like the noise meter app described on page 20, turn your smart phone into a measuring tool for collecting primary data. You can also use your smart phone to provide secondary data that may be useful to you while you are working in the field. There are several apps that you might be able to use. Figure 19 shows four of them and suggests how they might be used during fieldwork.

Activities

1 **Students wanted to investigate a woodland growing on a hillside. Their enquiry question was *'Are species of tree affected by altitude or angle of slope?.'* The most common trees in the woodland were oak, birch, ash, hawthorn, and holly.**

a) Select and justify a sampling strategy for this enquiry.

b) Select up to four apps that could be used and explain why you have chosen them

c) Design a data collection sheet for use with this enquiry.

Use a plant identification app to help you recognise plants. This might be useful if your enquiry investigated the effects of trampling in an ecosystem such as a sand dune or an urban park.

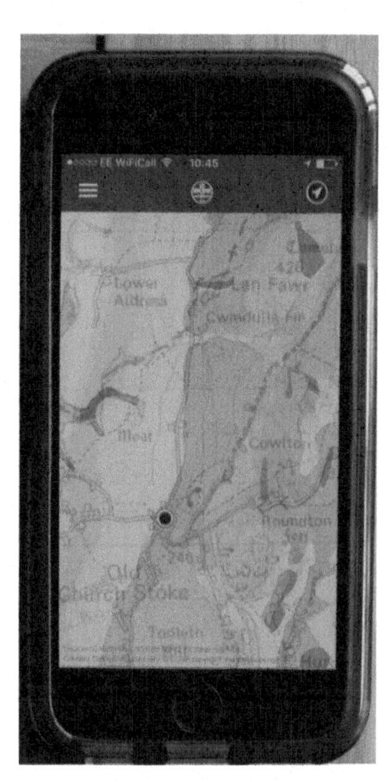

The iGeology app uses your phone's location to display a geological map of your location. You could use it to see whether local rocks are permeable or impermeable. This might help if your enquiry involves an investigation of flows of water through a drainage basin.

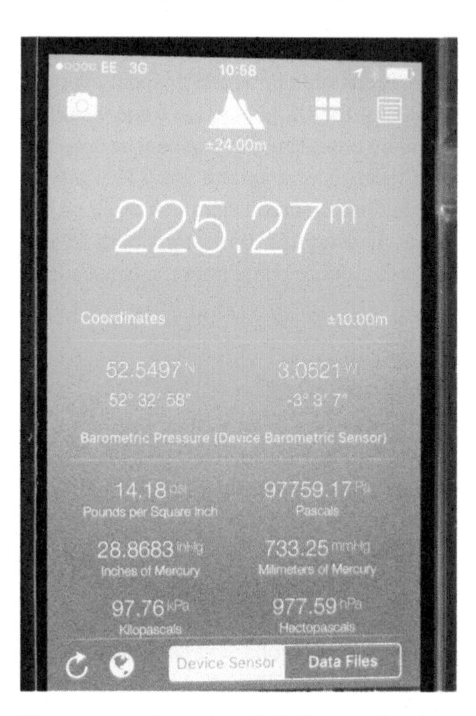

There are several apps that will display your elevation above sea level. This could be used to record your height if your transect went up and down a slope. It might also be useful if you were investigating whether land is at risk of flooding. This app will access your phone's camera so you can take images with a record of the time, location and elevation, as in Figure 17 on page 25.

Use an app like this to measure distances, for example, when collecting data along a longer transect (or long profile) through a large town or across a rural area.

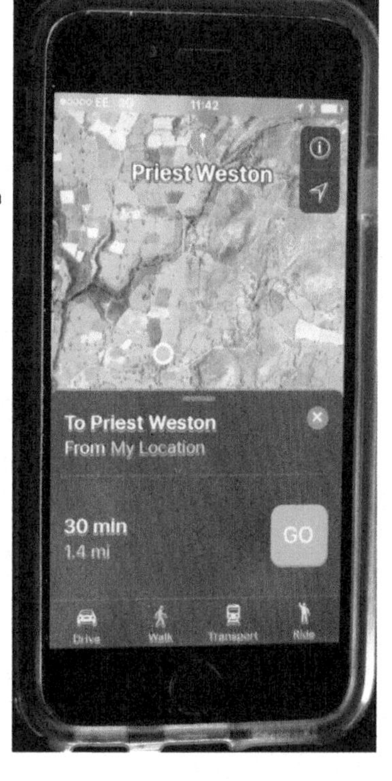

Figure 19 Smart phone apps that you might use during your fieldwork

Learning objectives

- Understanding the importance of big data.

https://www.riverlevels.uk/ This website publishes discharge data for rivers in the UK. You can use it to see how the discharge of a river changes with time. It is too dangerous to collect river velocity data when the river is in flood, so this website is particularly useful for seeing how river discharge changes after heavy rainfall.

https://www.nomisweb.co.uk/census/2011 The official UK National Census website

Big data

Some websites publish very large databases containing huge amounts of evidence – sometimes known as **big data**. These websites provide secondary data that may be of use to your enquiry. For example, you may be able to find a connection between primary data you have collected and secondary data published on a website:

- House prices (on a website) might be explained by variations in quality of the urban environment you have found using an EQI.
- Daily rainfall for the last month (on a website) might help you explain patterns of discharge you have found in a river and its tributary.

The National Census

Sometimes it can be difficult to answer the aims of your enquiry if you are only using primary data. For example, you may need to know such things as:

- the age structure of the population;
- the main occupations that people have;
- how many cars are owned by each household;
- the level of education that people have achieved.

You might be able to find this data through primary data collection by using a questionnaire. However, this would be very time consuming. Besides, people are sometimes reluctant to talk about their age, their education or their jobs so collecting this as primary data would be tricky. Thankfully, the National Census collects and stores this kind of evidence. It contains information about every household in the UK and includes data about the UK population's age, health, occupations, and education, amongst many other things. The census does have one major limitation: it is a snap-shot in time, recording data every 10 years. Your fieldwork could be located somewhere that has changed significantly since the last census. For example, an enquiry that investigates the sustainability of a new housing estate could not be supported by secondary data from the census if it was built after the census date.

Figure 20 Examples of enquiry questions that would need some secondary data

1 Are house prices higher closer to the park?
2 Is it healthier to live in the suburbs or in the city centre?
3 Are fewer crimes recorded in areas that have more CCTV cameras?
4 Are the residents of sustainable communities mainly in professional jobs?
5 Is car ownership higher in suburbs than in the city centre?

https://www.streetcheck.co.uk/

This website takes some of the data from the UK National Census and combines it with other big data sets such as data on crime and house prices. It can be searched using place names or postcodes. It provides numbers that you can process. It also presents some data in the form of graphs. You can search the database by postcode. A typical postcode unit (like CF5 2YX) has 200-300 people living in it. A postcode district (like CF5) can have between 8,000 and 85,000 people living in it, depending on whether the area covered by the postcode is rural or urban. It would be SMART to focus your fieldwork enquiry on a small number of postcode units rather than a postcode district which would be too large.

Strengths and limitations of using census data	
Some strengths	**Some limitations**
The census covers the whole of the UK. You can use it to compare data in one neighbourhood to another so you can analyse spatial patterns. The census is repeated every 10 years so you can find evidence of how a place is changing (for example, by comparing data from 2011 with 2001).	Some places change very quickly so the census may not accurately reflect what a place is like on the day of your fieldtrip.

Using census data to select a sample

You can use big data to research your fieldwork location before you visit. For example, you could use the census to decide on a suitable sample of local residents. A suitable sampling strategy for a questionnaire would be a stratified sample – where the number of people in each age group that you interview is in proportion to the actual population. You would do this by completing the following steps.

Step One Use a website with census data. Select the ward and find the population structure data. Figure 21 shows the age structure of Borth, a small seaside town in West Wales.

Step Two Calculate the percentage of people in each age category. For example, to calculate the percentage of people who are over 65 in Figure 21:

Divide 460 by 2,070 (the total) and multiply the answer by 100 = 22.22

This means that, if you questioned 100 people in total, you should interview 22 who are over 65.

Step Three Calculate the number of people to interview if your total sample size is less than 100. So, if you only want to interview 50 people in total, you multiply the answers from Step Two by 0.5 (because 50 is half of 100).

Age of residents	Number of residents in each age group
Over 65	460
30-64	1000
Under 30	610
Total population	2070

Figure 21 The age structure of Borth, West Wales. National Census (2011)

Activities

1 **Use the data in Figure 21.**
   a) Calculate how many people would be interviewed in each age category in a stratified sample of 100.
   b) Calculate how many people would be interviewed in each age category in a stratified sample of 60.

2 **Study Figure 20. It gives examples of enquiry questions which rely on at least some secondary data.**
   Use the street check website to identify at least one type of secondary data that you could use to help you answer each of these enquiry questions.

Chapter 3
Processing and presenting evidence

Learning objectives

- Knowing when and how to use mean, median, and mode.

What do we mean by average?

Processing your data will allow you to see what the evidence is telling you. One useful processing technique is to find the average. This is a measure of **central tendency**, in other words, the average tells you something about the central point in your dataset. There are three different ways to express an average. These are mean, median, and mode.

Mean

The arithmetic **mean** value of a dataset is found by:

Step One Adding up all the data to find the total.

Step Two Dividing the total value by the number of items of data.

Geographers often use the mean to find the average for discrete or continuous data. It is especially useful if data has been recorded several times and varies a little each time. For example, you might take five wind speed readings at each location (as in Figure 1) and find the mean. The wind may be gusty so each reading is different. By calculating the mean you can average out these variations. Small variations can also be caused by experimental error. For example, you might measure the time it takes for a dog biscuit to float downstream over a length of 10 metres. You might repeat the experiment 5 times and find each reading is slightly different because you didn't stop the stop watch promptly each time. Then you can take the mean to find an average time that reduces the effect of this experimental error.

Figure 1 A student has collected data on wind speeds at three locations on a hillside. The data is shown in the table (values in metres per second).

Site A – lower slope	4.2	3.9	3.7	3.8	4.1
Site B – middle slope	3.9	4.2	4.4	4.2	4.0
Site C – upper slope	4.8	5.1	4.7	4.8	5.2

Median

The **median** is the middle value when all your data is arranged in rank order. Median is a useful measure to use if you have some extreme values in your dataset. The median is unaffected by these extremes so will still give you a fair measure of the central point in your data. To find the median value:

Step One Put the data in rank order.

Step Two The median is the middle value in an odd number of items of data. In Figure 2, the median value in Group A is 5. If you have an even number of items in your dataset, the median is the half way point (mean) between the two middle values in your dataset. In Figure 2, to find the median value of Group B, add 6 to 12. Then divide the answer by 2.

Figure 2 A student has collected data on the length of commute for three groups of workers in an office in Cardiff. The data is shown in the table (values in miles)

Group A	2	2	3	5	5	5	12	23	26	
Group B	1	2	4	4	6	12	18	22	56	103
Group C	1	1	2	3	4	7	8	12	15	43

Section 1

Mode

The **mode** is the most frequently occurring value in your dataset. Mode is a useful measure of discrete data in a geographical enquiry. For example, you could use mode to describe the average value collected in a bipolar survey, as in Figure 3.

Step One Tally all the times someone chose a particular value in the bipolar survey.

Step Two The mode is the value that has been chosen most frequently.

	2	1	0	-1	-2	
Pavements are in good condition	2	3	5	18	12	Pavements are in poor condition
Cars never park on pavements	5	10	8	6	1	Cars often park on pavements

Figure 3 A bipolar survey was used as part of an enquiry about reducing risk for pedestrians and cyclists in a busy town. 30 people were surveyed. This table shows the total number of people who chose each value on the scale +2 to -2.

Mode can also be used to find a central point in grouped data. For example, imagine a beach survey in which pebbles are measured and put into categories by size, as in Figure 4. The **modal class** is the category that has most pebbles in it.

Pebble size (mm)	Number of pebbles
Less than 50	32
50-60	28
60-70	15
71-90	15
91-120	10

Figure 4 A student measured the length of 100 pebbles at one location on a beach. The data is shown in the table (values in millimetres).

Strengths and limitations of using averages	
Some strengths	**Some limitations**
Mean will give a number on a continuous scale. This is useful if you want to process the data further, for example, by adding the mean value to a scatter graph (see pages 48-51). Mode is always a data value and median will be a data value if you have an odd number of items of data. This is an advantage if you want to present findings that make common sense. For example, in a survey of car ownership, you may find that the mean number of cars owned per household is 1.86! It would be more sensible to report that the mode was 2 cars per household.	Mean can fail to represent the central point fairly if the dataset includes some extreme values or a very wide range of values. Median is a better measure if your data includes extremes.

Activities

1. **Calculate the mean wind speed at each location in Figure 1.**
2. **Study Figure 2.**
   a) Calculate the mean, median and mode for each group of commuters.
   b) Explain why the mean value does not fairly represent the central point of this data.
3. **Study Figure 3. What is the mode for each statement? What does this tell you about this survey?**
4. **Study Figure 4.**
   a) Which size of pebbles is the modal class?
   b) Identify two faults in the data collection sheet.

Learning objectives

- Knowing when and how to use range and interquartile range.

Measures of dispersion

Dispersion is a description of how your values are spread (or dispersed) through a dataset. The simplest measure of dispersion is **range**. This is the difference between the highest and lowest value in your dataset. This is useful because the range of one dataset may be very different to the range of another, even if the average value of the two datasets is similar. For example, in Figure 5, quadrats have been used to sample pebbles at two different locations on a beach. The table shows the length of 15 pebbles sampled from each quadrat. The range in values is quite different even though the median is the same. This suggests that wave action has moved pebbles up and down the beach and has sorted them by size – leaving pebbles with a very narrow range of sizes further up the beach.

Figure 5 Pebble sizes at two locations on Vik beach, Iceland

Length of pebbles (mm)	
Quadrat 1	Quadrat 2
95	25
60	22
35	20
30	20
20	18
18	15
14	14
14	14
11	12
10	11
9	10
9	9
8	9
8	7
7	8

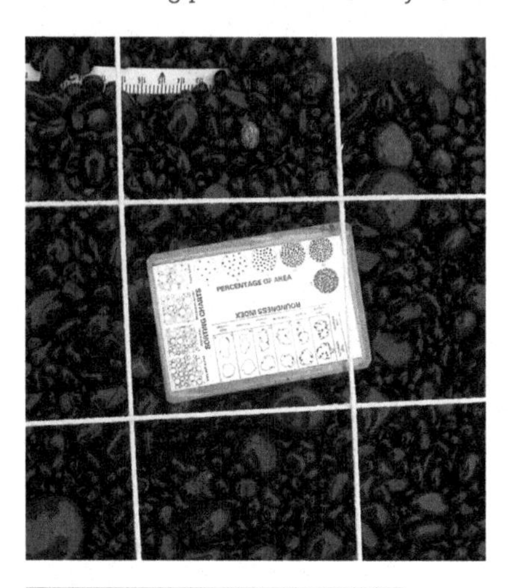

Figure 5.1 Quadrat 1. Pebbles close to the water's edge

Figure 5.2 Quadrat 2. Pebbles 5 metres from the water's edge

Interquartile range

The **interquartile range** is another way of measuring the spread of values in a dataset. The interquartile range is the difference between values that are one quarter and three quarters through the dataset. To find the interquartile range:

Step One Put the values into rank order and identify the median value. In Figure 5 it is in the green cell.

Step Two Identify the lower and upper interquartiles. The lower quartile is the mid-value between the lowest value in the dataset and the median. It is in the orange cell in Figure 5. The upper quartile is the mid-value between the median and the highest value and the dataset. It is in the yellow cell in Figure 5.

Step Three Calculate the range by subtracting the value of the lower quartile from the value of the upper quartile.

Dispersion graphs

The range of values in a dataset can be presented visually using a **dispersion graph**, like Figure 6. Figure 6 shows the range in house prices in Borth, a seaside town. To make a dispersion graph:

Step One Set out your categories of data on the horizontal axis. In Figure 6, these are different locations within Borth.

Step Two Mark the vertical axis with a range of values for your data. In Figure 6, this is house prices. The highest value is £380,000 so the axis is marked at £50,000 intervals up to £400,000.

Step Three Plot each value in your dataset as a small vertical cross. In this case, each cross represents the value of one house that is for sale in Borth.

Figure 6.1 Borth in West Wales

House price (£)

400,000
350,000
300,000
250,000
200,000
150,000
100,000
50,000
0

At sea level
Above sea level

Figure 6.2 Dispersion graph showing range of house prices in Borth (2017)

Activities

1 Study Figure 5.

a) For each quadrat, calculate the:

i) range ii) median iii) interquartile range

b) What conclusions can you draw from these results?

2 Study Figure 6.2.

a) What do the red crosses on this graph represent?

b) Use the graph to calculate the interquartile range in house prices for these two locations.

Types of graph

Learning objectives

- How to choose the right kind of graph for discrete and continuous data.
- How to present data using a line graph.

Graphs are a way of presenting your data in a visual form. They make it easier to see patterns or trends within your data. They also make it possible to compare data easily.

There are lots of different types of graph you can draw but choosing which graph to draw is not just about personal taste. There are some rules about which graphs you can use.

Two of the most common graphs are:

- **Line graphs** which are used to present continuous data and can be used for discrete data.
- **Bar charts** (or column graphs) which are used to present discrete data and must **not** be used for continuous data.

Bar charts and line graphs each have two axes. The horizontal axis is known as the x-axis and the vertical axis is the y-axis.

Line graphs

Line graphs are used to present continuous data. **Continuous data** is data that is measurable and the result may be recorded to one or more decimal places. Distance, height, slope angle, sound level, velocity (for example, flow of a river or wind speed) and time are all examples of continuous data collected in fieldwork. Figures 9 and 10 each show continuous data displayed as line graphs.

Discrete data in one category, such as the amount of traffic, can be shown as a line graph if the x-axis shows time (for example, in hours, days, or years). Figure 7 shows an example. However, you should never present separate categories of data, such as different shop types, joined by a line.

Time	Pedestrians
8:00	36
8:30	78
9:00	69
9:30	43
10:00	44
10:30	53
11:00	57

Figure 7: Pedestrians in Cardiff Students collected data on pedestrian flows in a busy shopping street. They counted the number of pedestrians who passed by in three minutes every half an hour.

The x-axis of a line graph can be used to show distance along a transect. For example, you can present the slopes of a beach profile as a line graph (see pages 16-17). Figure 10 shows the height of plants growing on a transect that cuts at right angles across a footpath. Cross sections, such as the shape of a river channel, can also be presented as a line graph.

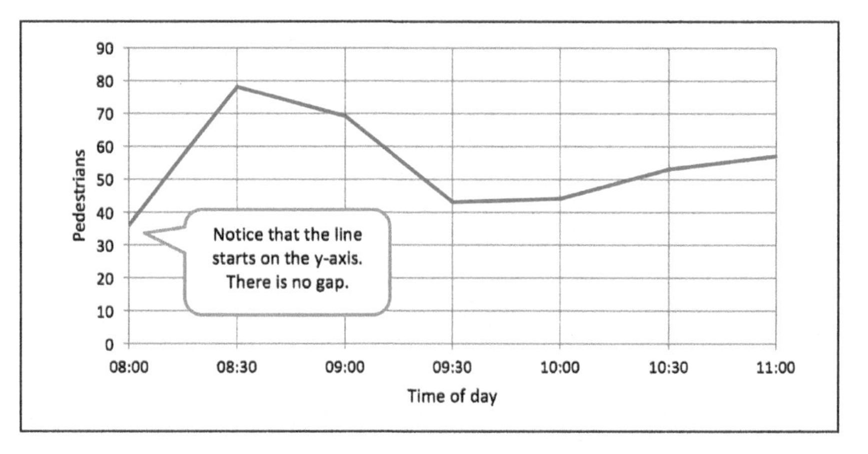

Activities

1. **Discuss why it is important to start a line graph on the y-axis.**
2. **Suggest one strength and one limitation of using the y-axis to represent height in Figure 9.**

Figure 8 Students collected data on flows in this rural honeypot site. They set up one transect up the west facing slope, from top to bottom. They collected wind speed data. They also set up a short transect across the footpath to see how plants were affected by trampling. Shropshire Hills, AONB

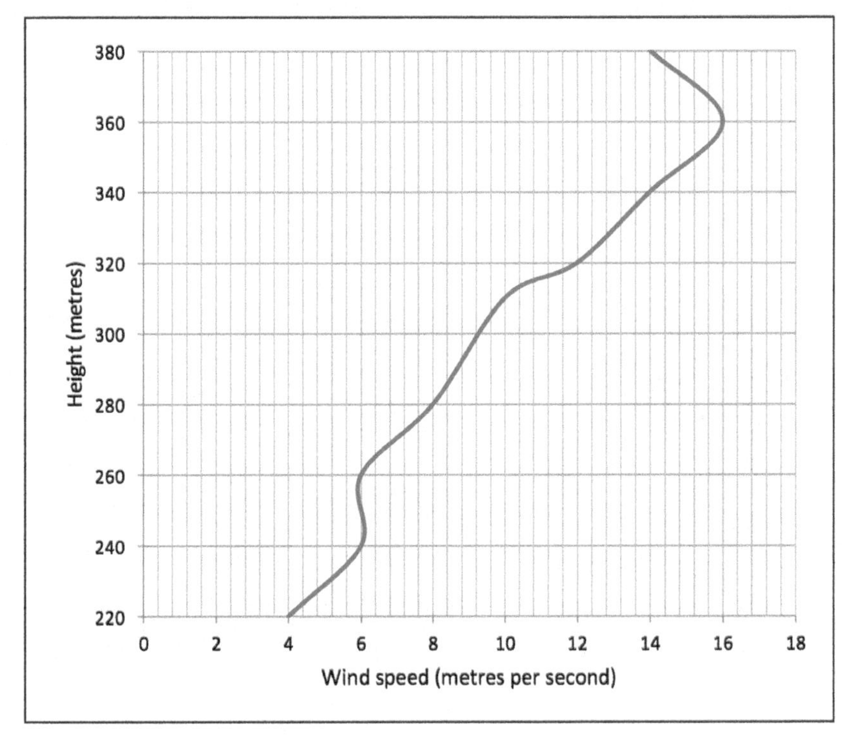

Figure 9: Wind speeds
Line graph of wind speeds. Notice that the graph shows height of the hill on the y-axis.

Figure 10: Plant height
Cross section across the footpath showing plant height

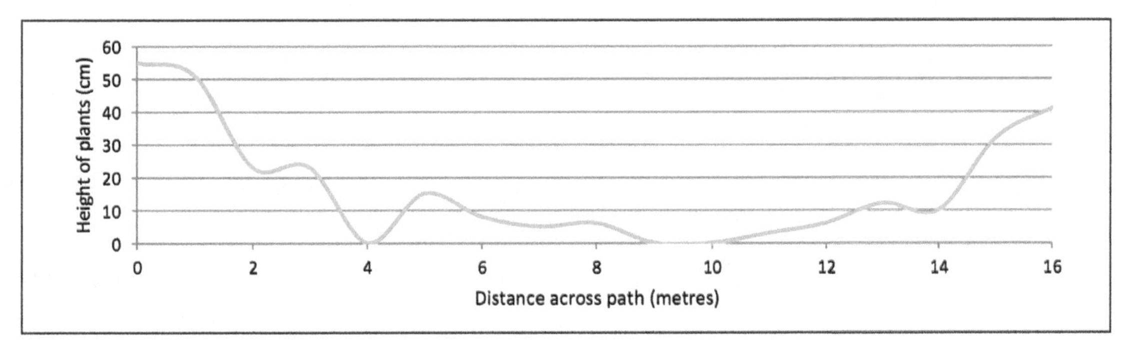

Bar charts

You should choose to use a bar chart if you have discrete data. Discrete data is data that can be counted and the result is a whole number. The number of pedestrians in a shopping area and the number of cars passing in 5 minutes are both examples of discrete data. The number of items in a category, for example, the number of pebbles of a particular shape, is another example of discrete data. If the bars are drawn vertically as columns, then the x-axis is used for the categories of data and the y-axis is used for the values. For example, traffic can be shown as vertical bars with different categories of vehicle on the x-axis. Bar charts clearly show the category with the largest and smallest values so are useful for comparing your categories.

Hints and tips for creating bar charts

- It is possible to draw horizontal rather than vertical bars. There is no rule to decide whether your bars should be vertical or horizontal but if your category names are very long then it is easier to label horizontal bars.
- Some categories have a logical sequence and should be shown in the correct order, like Figure 12. However, if your categories have no particular sequence then you can put your data in rank order by value from largest to smallest – making comparison easier. An example is shown in Figure 13.
- Decrease and increase can be shown by using positive and negative numbers on either the x-axis (like Figure 14) or the y-axis (if you want a column graph).

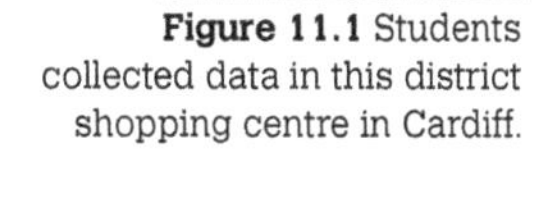

Figure 11.1 Students collected data in this district shopping centre in Cardiff.

Figure 11.2 Students recorded shop type on each side of the street. They compared their results to images on Google Street View in 2008 (secondary data).

Shops in Cowbridge Road East	2008	2016
Food shops	2	6
Newsagents	1	1
Cafes/pubs/take aways	20	26
Estate agents	2	3
Health care/chemists	6	2
Furniture / home ware	5	3
Clothes	5	4
Electronics	1	1
Hair salon/beauty	2	2
Charity shops / betting / amusements	10	18
Vacant shops	9	2

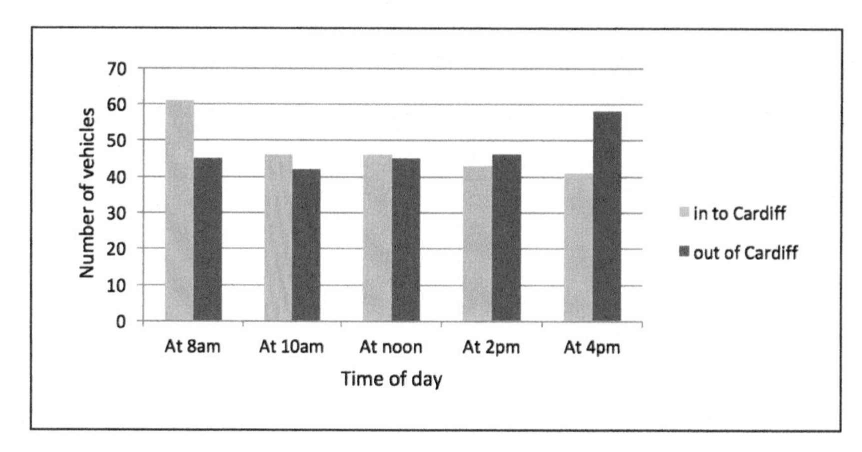

Figure 12 Traffic data was collected in the street shown in Figure 11.1. Vehicles were recorded travelling in each direction for 3 minutes. The results are presented in a vertical bar (or column) chart

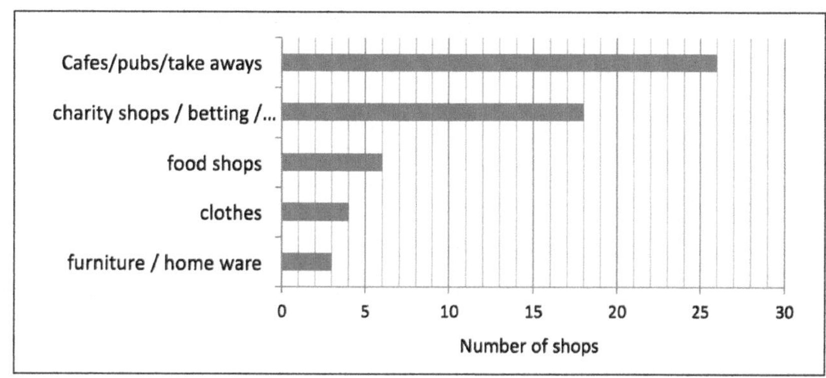

Figure 13 The five most common shop types in 2016 are presented in this bar chart

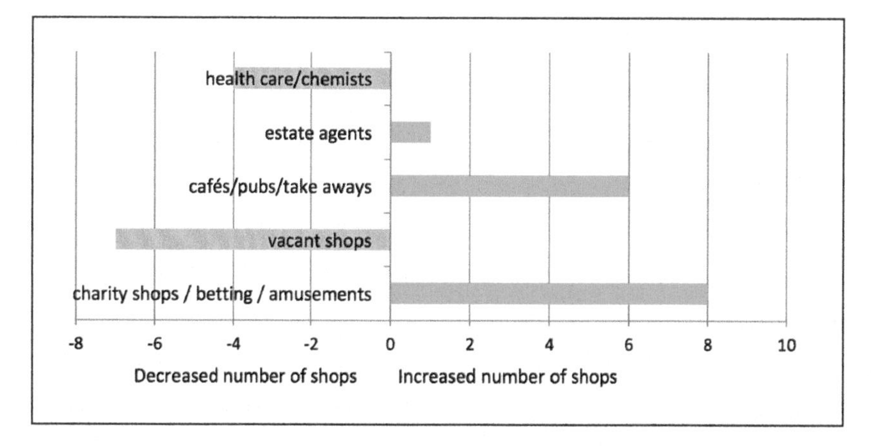

Figure 14 The change in the number of shops in selected categories between 2008 and 2016 is shown in a bar chart that has an x-axis with both positive and negative numbers

Activities

1 **Study Figure 12.**
   a) Explain why a bar chart has been selected to present this data.
   b) What conclusions can you reach from this data?
2 **Use the data in Figure 11.2 to practise drawing:**
   a) a horizontal bar chart with the data in rank order;
   b) a vertical bar chart showing the change from 2008 to 2016.

Learning objectives

- How to avoid common mistakes when drawing graphs.

Graphs: getting the details right

It is important to get the details right so that people who view your graph can fully understand what it is telling them about the data. A good graph, selected carefully to make sure it is appropriate for the data (discrete or continuous), will have:

- a Title;
- Labels on both the x-axis and y-axis;
- a Key, if you have used colours or symbols;
- been drawn neatly and accurately with a ruler.

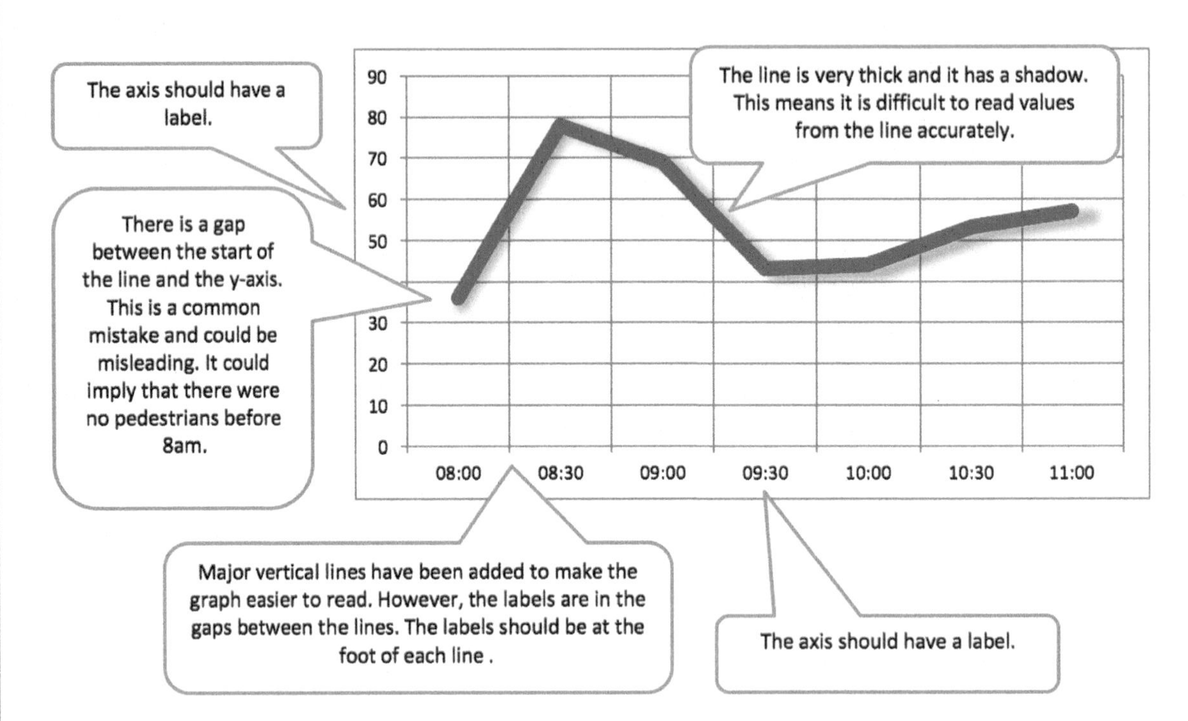

Figure 15 This graph contains some errors. Compare it to Figure 7 on page 32

Dos and don'ts of drawing graphs

Do:	Don't:
✓ Label each axis and check that you have included the unit of measurement (for example, cm or m/s) in your label; ✓ Start the origin of the y-axis at zero unless you have a good reason not to.	✗ Include a gap between the start of a line graph and the y-axis.

Misleading graphs

A good graph allows you to see the data clearly. If the graph is badly drawn it could be misleading – making it difficult to see patterns or make comparisons. One common mistake is to provide a false origin for the y-axis – in other words, one that does not start at zero. Study Figure 16. This pair of graphs shows what happens if the y-axis does not start at zero. Students had collected data on pedestrian flows in a busy shopping street. The left hand graph gives the impression that more than twice as many pedestrians were seen at site B than at site C. However, the y-axis starts at 60 not zero, so differences in height between the bars has been exaggerated. In the right hand graph the y-axis starts at zero. You can see that site A has more pedestrians than B, and B has more than C. However, the differences are actually quite small.

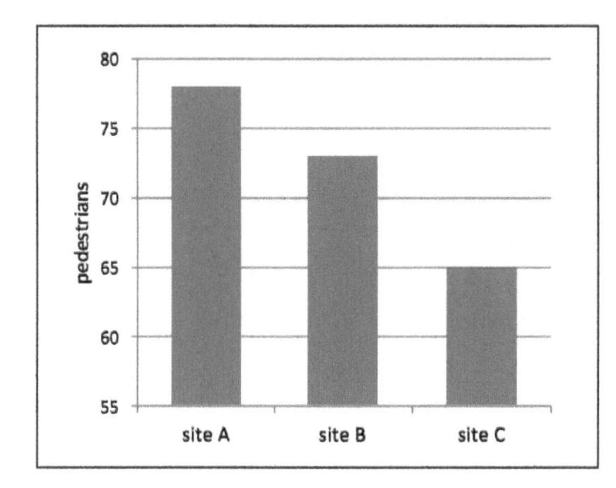

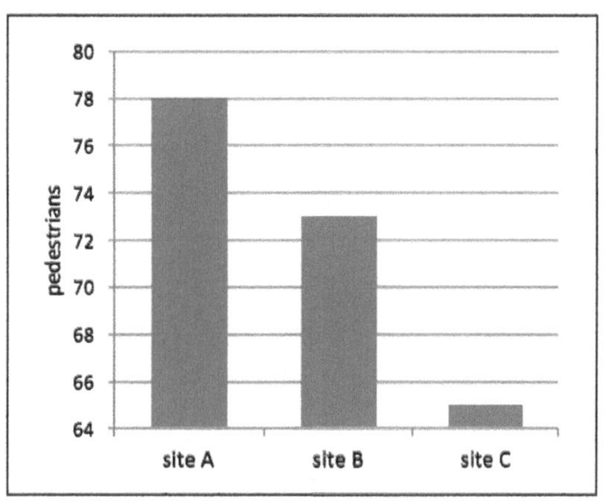

Figure 16 Misleading bar charts

Pictograms

Pictograms are simple pictures that are used to represent quantities. The picture is drawn in proportion to the quantity it represents. Study Figure 17. On the left the pictogram is drawn correctly. Each pictogram represents 10 cars, so location A had 45 cars. On the right the pictogram is drawn incorrectly. The value is 3 times greater so the car has been enlarged until it is 3 times its original length. However, this is misleading. The car is also 3 times its original height. This means the pictogram is actually nine times larger than the original pictogram, suggesting that it represents 90 cars rather than 30.

Figure 17 Misleading pictograms

Learning objectives

- Understanding the importance of annotation.

Annotating images

Annotation means to add analysis to an image. Annotation is more complex than labelling. Labels are simple descriptions. Annotations should be used to analyse key features of the image. The analysis should demonstrate your understanding of those features based on the evidence you have collected during fieldwork. Figure 18 gives an example. Annotation has been added to the sketch map of Cardiff which was described on page 23.

Figure 18 An annotated sketch map of a fieldwork location. Central Square, Cardiff

Noise carried further along Transect 1 because of the wind direction. Sound was also able to travel along the river because there are no large buildings to block it.

Noise from the construction site in Central Square was not carried far along this transect because it seemed to be blocked by the tall buildings. Also, noise from this busy road masked noises from the building site.

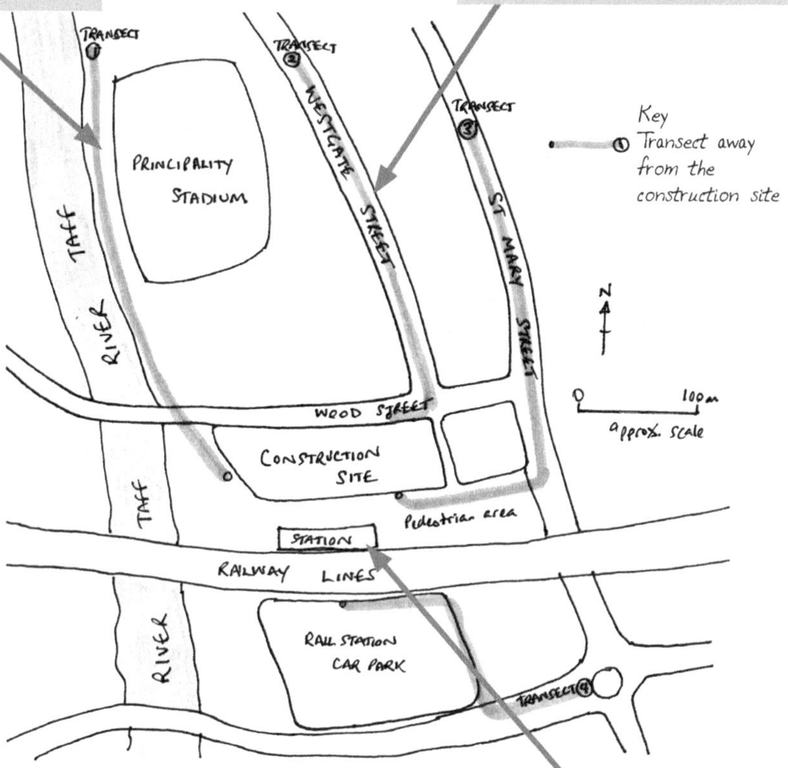

The wind was blowing from the south east. This meant that sound from the building site was carried by the wind further to the west and north west than to the east or north east.

Noise from Central Square was largely blocked by the railway station.

Strengths and limitations of annotation	
Some strengths	**Some limitations**
Annotation should help you to pick out and analyse the key features of an image quickly and precisely. Annotation can be used to analyse a wide variety of images including photographs, field sketches and maps.	You may have a limited amount of space for your annotation so choose your words carefully.

Drawing maps

Geographers are interested in spatial patterns. These are patterns that exist in two dimensions so they can be represented on a map. Can you represent some of the evidence from your fieldwork on a map? You can use either primary or secondary data. If so, you will need to select the most suitable type of map. Figure 19 will help you select a suitable technique.

Learning objectives

- How to choose which type of map to draw.
- How to draw isoline maps and located bar maps.

Figure 19 Which type of map?

Type of data	Examples of this type of data	Suitable types of map	Strengths (+) weaknesses (-)
Data has been collected for two or more areas (for example, wards in a town).	Data from the census such as the percentage of people who have a specific level of education or category of job or live in rented accommodation.	**Choropleth** maps. Each area on the map is shaded with a colour that represents a range of values.	+ Bolder colours or denser shading are a visual way to represent larger values. - Too many or too few colours and it will be difficult to see any patterns.
Where the same type of data has been collected at many points.	Discrete data (for example, traffic counts) or continuous data (for example, noise levels) that have been collected at many points across an urban area (perhaps simultaneously).	**Isoline maps**. Each line on the map represents an equal value.	+ Contour maps are a type of isoline map so these maps are great for showing surfaces and gradients, for example, how a good (or bad) impact declines with distance in a sphere of influence study. - You will need a lot of data points.
Where two or more items of data that you would like to compare have been collected at a few points.	Qualitative data from a bipolar survey (or Likert Survey) for two or more variables, for example. scores for noise nuisance and litter in an investigation of sustainability.	Proportional symbols (such as bars) that are located on a map.	+ The bars are easy to draw and make comparisons between locations easy. - Long bars will start in the correct location but end elsewhere on the map. This may make the map difficult to analyse.

Located bars

Bars are good for comparing discrete data (see pages 34-35). By locating bars on a map, you can compare the values at different locations.

Step One Draw a simple base map which includes the points where data was collected.

Step Two Create a vertical scale for your bars. This needs to be small enough so that your largest bar will fit on the map neatly but not so small that the bars are fiddly to draw. You might do this on graph paper.

Step Three Draw each pair of bars. You could do this directly on the map, or you could draw them on graph paper and glue them to the map. Position the foot of each bar as close as possible to the place where data was collected.

Step Four Add a scale line, north arrow, and key.

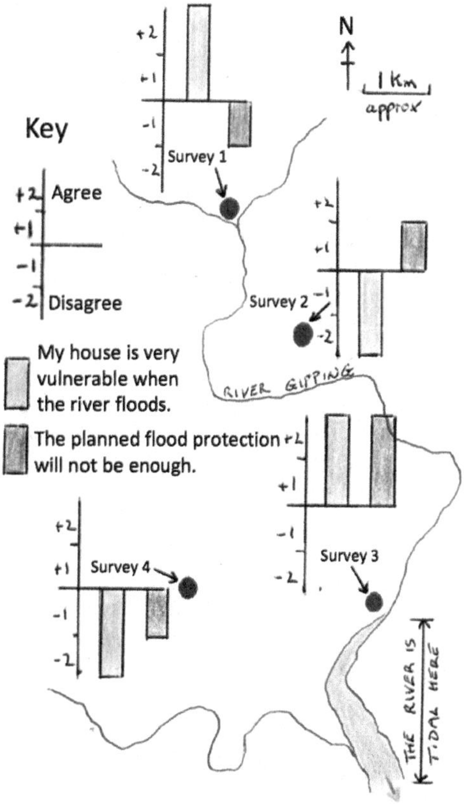

Figure 20 Located bars map showing different perceptions of flood risk

Learning objectives

- How to draw isoline maps and flow line maps.

Flow line maps

A **flow line map** is a suitable presentation technique to show flows or movement between places. The width of each arrow drawn on the map is in proportion to the value of the flows. For example, you can use proportional arrows to represent flows of traffic or the movement of pedestrians. To construct a flow line map, complete the following steps.

Step One Draw a simple base map which includes the points where data was collected.

Step Two Create a scale for the width of your arrows. This needs to be small enough so that your largest arrows will fit on the map.

Step Three Draw each arrow onto the base map so that they point in the direction of flow. Position each arrow as closely as possible to the place where data was collected.

Step Four Add a scale line, north arrow, and key to your map.

Strengths and limitations of flow line maps	
Some strengths	**Some limitations**
Flow lines are a great visual way to represent both amount and direction of movement. They are easy to draw using ICT. Use software such as Power Point and insert arrows. These can be dragged to the correct width and direction.	Arrows need to be drawn carefully. They will be confusing if they overlap or cross each other.

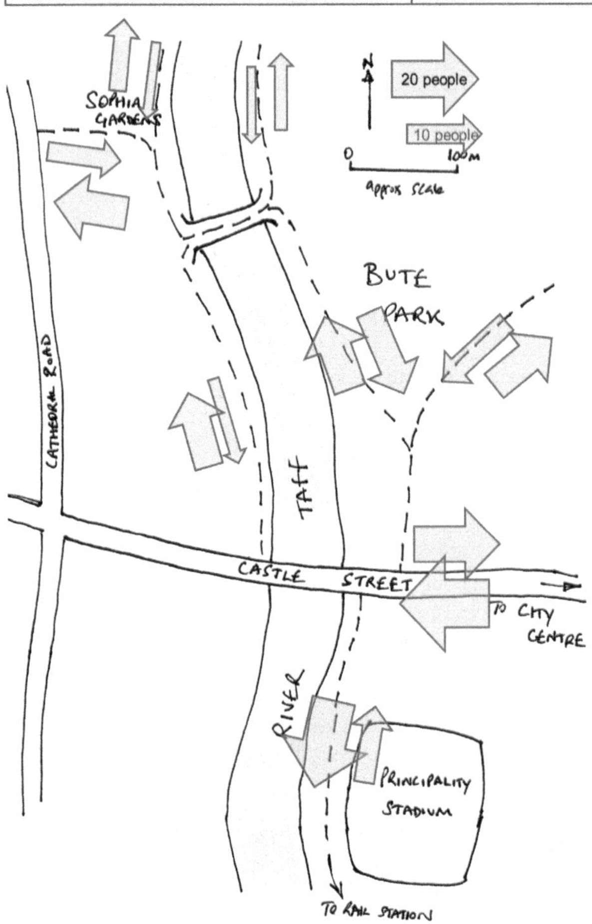

Figure 21 A flow line map representing pedestrian flows

Isoline maps

An isoline is a line on a map that joins places that have an equal value. In fieldwork about spheres of influence you could create an **isoline map** to show noise levels around a building site, stadium, or busy road.

To draw an isoline map you will need lots of points across a map where data has been collected. Once you've got enough data you can draw the map – a bit like a dot to dot map. However, you also need to be able to interpolate - which means to find values that lie between other values.

Study Figures 22 and 23. Each dot on Figure 22 represents a place where data has been collected. The students visited each place and gave a score (out of 5) for the amount of litter that they could see. A value of 5 means lots of litter (either in a bin or on the street) and 0 means there was no litter.

Study Figure 22. There are 5 places where a value of 2 was recorded. We could join them up but it would be easier if there were more points. To create more points we need to interpolate. Each of the red dotted lines has been drawn between values where 2 should be somewhere on the line. Using this method gives us another 5 places that have a value of 2.

Figure 23 shows how to interpolate the data and draw an isoline for 2. Everywhere inside this line has more litter. Everywhere outside this isoline has less litter – it effectively shows a sphere of influence.

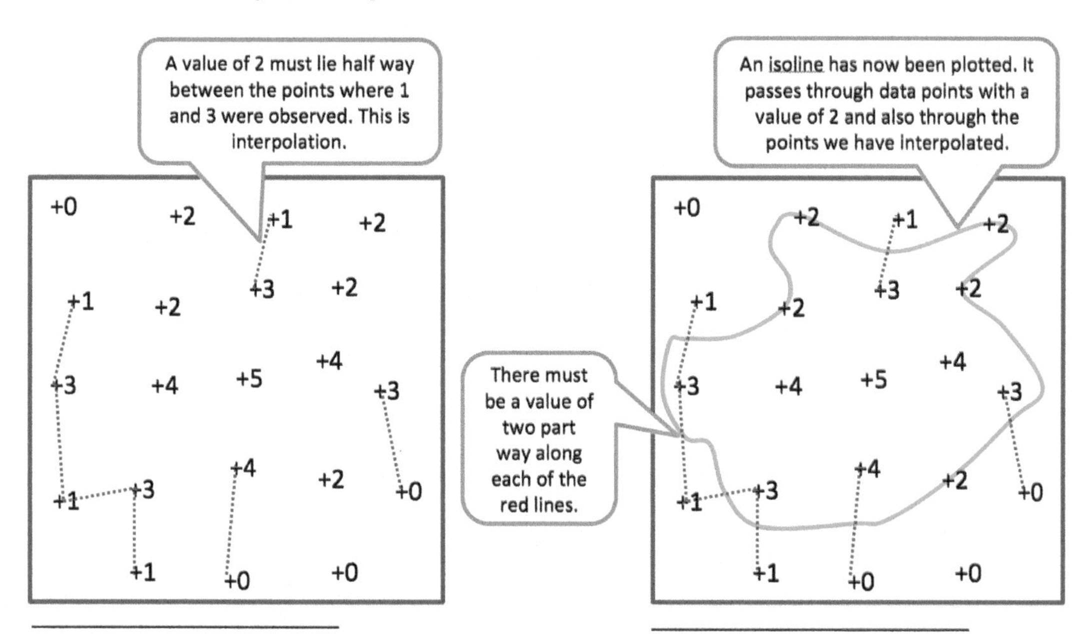

Figure 22 Interpolation of extra data points with a value of 2

Figure 23 An isoline with a value of 2 has now been plotted

Activities

1 Make a copy of the data points in Figure 22. Use the interpolation technique to draw an isoline for the value of 4.

Chapter 4

How can evidence be analysed?

Learning objectives

- How to analyse trends on a graph

What is analysis?

Analysis is like detective work. You have collected a lot of data. Now you need to sift through this data carefully, selecting the evidence that reveals connections, patterns, or trends. This process will allow you to make sense of the data.

Like a detective, you will need to examine a variety of evidence. You may have collected evidence in the form of tables of data, photographs, interviews, graphs, charts, or maps. The techniques that will help you make sense of this evidence are described in this chapter.

As you begin to analyse your data it might be helpful to ask yourself some questions. For example:

- ***Do I have any evidence for ...?***
- ***Why do I think that...?***
- ***How clear is the pattern or trend?***
- ***How strong is the connection between this variable and that?***
- ***What may be causing this variation in the data?***
- ***What effect does this variable seem to have?***

Three key points

A simple technique that you can use with almost any kind of evidence is to identify the three key points that the data is telling you. If you are making sense of a graph or a table of data you might look for:

- the overall trend – is it going up or down?
- the range – the difference between the highest and lowest values;
- any values that don't fit the overall trend. We call these anomalies.

https://www.riverlevels.uk/esk

Figure 1 shows an example of a hydrograph. This is a line graph that shows how the amount of water in a river changes over time. If you want to challenge yourself:

- try to identify five key points instead of three;
- use descriptive words that describe the patterns or trends you can see. Figure 2 gives examples of useful words to use.

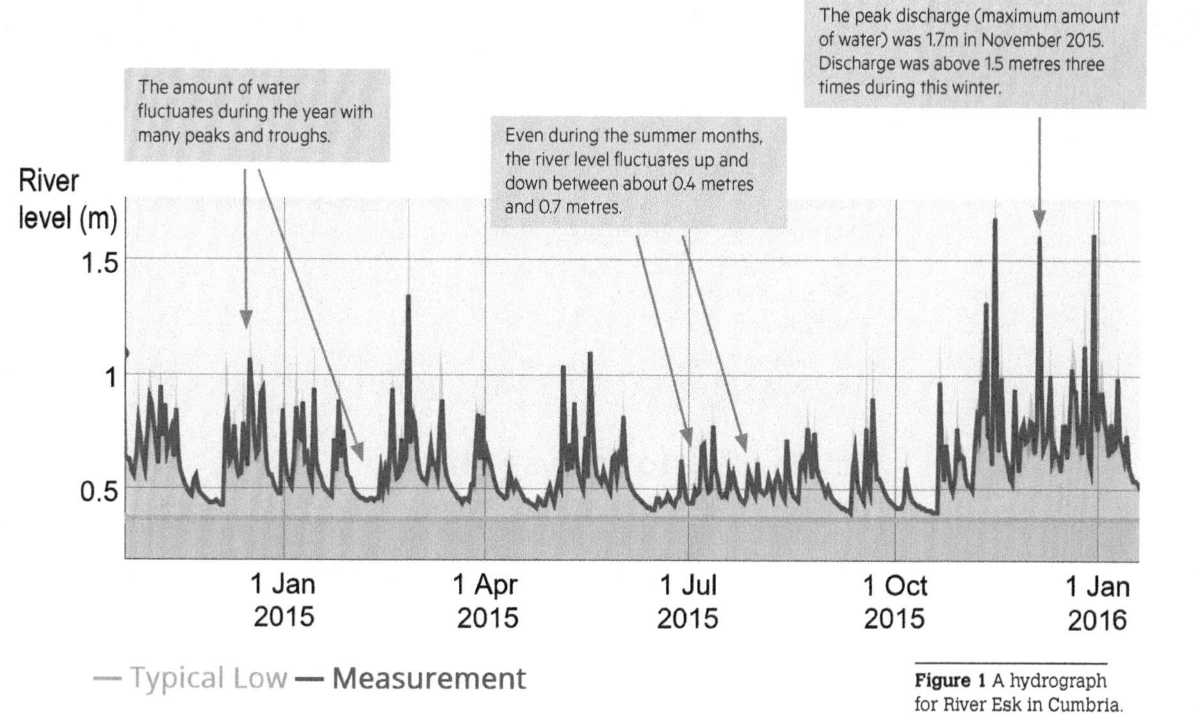

Figure 1 A hydrograph for River Esk in Cumbria. Three key points have been identified.

Clustered	Values (or points on a map) are concentrated into small groups.	**Linear**	Features (on a map or photograph) are spread out along lines.
Decreasing	The values are going down. Add **slowly**, **steadily** or **rapidly** to describe the trend.	**Random**	Values (or features on a map or photograph) are at irregular spaces and show no clear pattern.
Fluctuating	The values in the data vary up and down in a repeated pattern.	**Regular**	Values (or features on a map or photograph) are repeated at even spaces.
Increasing	The values are going up. Add **slowly**, **steadily** or **rapidly** to describe the trend.		

Figure 2 Useful descriptive words to make sense of data

Activities

1 Add two more key points to Figure 1.

Text analysis

Learning objectives

- Understanding simple ways to evaluate text.

Text is an important source of evidence. It can provide both facts and opinions. The way that the text is written may tell you something about what somebody thinks of a geographical issue – their perception. You may have collected text as part of your evidence if you have interviewed people or asked open questions in a questionnaire. You may also have collected text evidence from books, newspapers, or the internet in the form of:

- blogs;
- interviews;
- adverts;
- news articles;
- reviews.

Simple ways to analyse text

How do you analyse any patterns or trends when your evidence is in words? There are a number of simple things you can do.

1. Use one colour of highlighter to identify facts and another colour to identify opinions. In Figure 4, one fact has been highlighted in yellow in interview 1. One opinion is highlighted in green.
2. Consider the tone of the text – what kinds of words are used? Are the words mainly positive or is a negative tone used in some of the text? You can count the words or phrases that have a positive tone and then compare those words/phrases that have a negative tone in a bar chart. In Figure 4, one positive phrase has been highlighted in pink in interview 2. One negative phrase is highlighted in blue.
3. Copy and paste your text into an online tool that counts the frequency of words that are used. The text of the interviews in Figure 4 has been analysed in this way and the top 20 words used in each interview are shown in Figure 5. You can analyse these lists. Are the words mainly nouns that identify physical features of the environment, or human features? Are there many adjectives? The number of times that adjectives such as 'beautiful', 'pretty' and 'friendly' can be graphed using a bar chart. This will tell you something of the positive (or negative) perception of a place.

Activities

1. **Analyse the text in Figure 4.**
   a) Identify one fact and one opinion in each interview.
   b) Count the number of times that each interview describes:
      i) a physical feature of the environment;
      ii) a human feature of the environment.
   c) What are the main similarities and differences between Interviews 1 and 2? Use evidence from Figures 4 and 5. You could draw bar charts to help this analysis.
2. **Describe how the interviews in Figure 4 could be used as a pilot survey to create a more closed qualitative survey. See page 19 to help your answer.**

Figure 3 Borth, a small seaside town in west Wales.

Figure 4 Transcripts of two interviews. Students were investigating the concept of place in Borth. They interviewed visitors and recorded what they said about the place

Interview 1 Borth is a small seaside town in West Wales. It is a wild and natural place that is close to nature. The estuary is constantly changing: the light shines on the water and reflects off pools of water on the beach. Birds soar in the sky above. This place is magical. It's as though the landscape is alive. The light is beautiful. I think it's because the weather changes so much and we have light that reflects off the estuary and the waves. The beach is glorious. When the tide is out a long way the wet sand glistens and shines in the light. The sand is soft and clean. I think the beach is really beautiful. I love walking along the pebbles looking for things that have been washed up in the tide. The beach is a lovely place but it can be wild in the winter when the storm waves crash against the pebbles. Then you know that nature can have wild forces.

Interview 2 Borth is a friendly place. Local people are very friendly and want to help. I like the main street with its cheap cafes and shops selling holiday souvenirs. You can get most things in the local shops but not everything so we have to drive quite a long way to do some shopping - like clothes. The main street has pretty little cottages on either side. Some of the old cottages are made out of pebbles off the beach. The town has a lot of interesting history. There are special legends about the place too - legends about the sea and the drowned forest. I think the history of the place is very interesting. Borth is pretty in the summer and the beach is beautiful but it can be rather bleak here in the winter and sometimes I feel quite isolated.

Interview 1		Interview 2	
Primary words	Frequency	Primary words	Frequency
light	4	place	3
beach	4	legends	2
place	3	history	2
wild	3	beach	2
pebbles	2	cottages	2
sand	2	pretty	2
beautiful	2	shops	2
water	2	main	2
nature	2	Borth	2
estuary	2	friendly	2
shines	2	local	2
reflects	2	very	2
think	2	street	2
waves	2	quite	2
tide	2	interesting	2
glorious	1	isolated	1
much	1	sometimes	1
weather	1	old	1
landscape	1	beautiful	1
magical	1	cafes	1

Figure 5 Text analysis of the two interviews to identify the most frequently used words

http://www.textfixer.com/tools/online-word-counter.php

Further ways to analyse text

If you can turn text into something else it can help you to analyse it and make sense of it. One simple way to represent text in another form is to count the key words and represent them in a bar chart – as suggested on page 44. Two other ways to represent text are shown in Figures 7 and 8. Both represent the text shown in Figure 6 below.

Figure 6 Text from the Visit Wales website describing Borth. Source: http://www.visitwales.com

> Borth's beach is three miles of gently shelving golden sand and is especially popular with families with younger children and sailboard enthusiasts. The tide goes out a long way, so its shallow waters are great for the little ones to paddle in and splash about.
>
> Dog restriction on section of beach 1 May – 30 Sept.
>
> At the southern end of this wonderful beach, an ancient submerged forest is exposed by the ebbing tide. Welsh legend has it that the trunks and tree stumps of old forests, long hidden under sand and sea, are the remains of the land of Cantre'r Gwaelod, which disappeared under the waves of Cardigan Bay, long ago.
>
> Cantre'r Gwaelod was protected by dykes and dams but one night a feast was held and the place's night watchman called Seithennyn became inebriated and neglected his duty, resulting in the submerging of Cantre'r Gwaelod. After a storm and the ebbing of a very high tide, more and more of these ancient tree stumps come into view.
>
> Blue Flag and Seaside Award beach. Toilets, cafes, restaurants, pubs, shops, parking. Lifeguard service provided 2 July to 4 September. Dog restrictions apply May to end September.

Word clouds

One way of representing text in another form is to create a **word cloud** (or wordle). You could use this type of analysis to explore how the identity of a place is perceived. A word cloud (or wordle) gives a very visual way of recording how people see and identify a place. The size and position of each word in the 'cloud' is in proportion to its importance in the original text. For example, in Figure 7 (which has been created using the text in Figure 6), the word 'beach' is large and in the centre of the word cloud because it is the most commonly used word in the original text passage. This enables you to see what people think about the place and how those thoughts shape their perception of identity.

Figure 7 A word cloud based on the text in Figure 6

A word cloud can be made by following these steps:

Step One Decide on an interview question that matches the focus for your investigation, or select text from an internet site.

Step Two Interview your target group (this might be related to age / occupation / gender). The more people you ask the better.

Step Three Decide if the word cloud is designed to represent images of identity before a visit / after a visit (or both if the aim is to compare images and identities before and after visits).

Step Four Input the text from all of those interviewed into a suitable website to produce the word cloud. The website allows you to change font and colour variation to add further visual impact.

Several websites allow you to create word clouds. Here are two:
http://www.wordle.net/
http://www.wordclouds.com/

Create spider diagrams

Word clouds are visual and simple to make. However, making a spider diagram to represent your text is a more sophisticated way to analyse the evidence. You can still pick out key features of the text by placing them in the centre of your diagram. The beach is a key feature of Figure 6 so it is central in Figure 8. The main advantage of using a spider diagram is that you can use it to identify the connections that bring meaning to the text. For example, Figure 8 has identified the physical and human features that make Borth special. These are picked out in the red clouds. It then uses connections to explain why these features are important. The most important conclusions of this analysis have been highlighted in orange.

Figure 8 A spider diagram based on the text in Figure 7

Activities

1. **Use a search engine to find a description of Borth. Choose a website that either encourages people to visit or reviews a visit to Borth.**
   a) Copy the text into an online text analysis tool and graph the results.
   b) Create a word cloud for your text.
   c) Analyse it using a spider diagram.
2. **Which of these analysis techniques seems to be the most useful? Justify your choice.**

Learning objectives

- How to draw a scatter graph

Making connections

To make sense of your data you also need to make connections between the evidence. Looking for evidence of cause and effect will help you to explain the evidence. For example, imagine an investigation of the shopping centre shown in Figure 9. You might investigate the patterns of pedestrian flows. Why are some parts of the town centre busier than others? You could collect evidence such as:

- pedestrian surveys at different times of day;
- interviews with shoppers about the features of the town centre that they like and dislike;
- the rateable value of shops could be found on the internet.

To make sense of this evidence you might look for connections such as:

- whether the busiest streets also have a lot of features that the shoppers like;
- whether rateable values are higher in the busiest streets. If so, this might explain why national chain stores are located on busy high streets but local shops are on quieter side streets.

Figure 9 Why is this shopping street particularly busy? Shrewsbury.

Activities

1 **(a) Suggest three different reasons why some shopping streets have more pedestrians (footfall) than others.**

(b) Identify the data that you would need to collect to investigate the reasons for variations in footfall in a busy shopping street.

Scatter graphs

A **scatter graph** is a type of graph that is used to show the possible connection (or correlation) between two sets of data. To draw a scatter graph you will need **bivariate data**. That is two sets of data (or variables) that you think may be connected in some way.

Step One Think about how the two variables may be connected. For example, in Figure 10, a student has collected data from the Census about the educational background and occupations of people living in 14 different neighbourhoods in Cardiff. The student thinks it is likely that more people will have professional occupations in wards where more people have a degree. If so, the percentage of people in professional occupations will be the **dependent variable** because the value of this number will depend on the percentage of people who have degrees – not the other way around.

Step Two Draw a pair of axes. The dependent variable should go on the vertical axis. The **independent variable** – in this case the percentage of people with a degree – should go on the horizontal axis. Try to make the two axes about the same length as each other so that the finished graph is square. This will make it easier to see any patterns on the graph.

Step Three Label each axis.

Step Four Plot the points for each ward onto the graph. Vertical crosses are the best way to plot each point. These will line up with the vertical and horizontal grid lines on your graph paper – so your plots should be accurate and easy to read.

Step Five Calculate the **mean (M)** value for each variable. Plot this point onto your graph – perhaps using a different style or colour so that it stands out from the other points on the graph.

Step Six Draw a line of best fit on your graph. The line must pass through the mean value (M). It should also follow the trend of the other points – but it doesn't need to go through any of them or join them up. However, there should be the same number of points on each side of the line of best fit. Notice that the line of best fit does not have to go through the origin of the graph.

Ward	% of residents with a degree	% of residents in professional occupations
Adamsdown	28	14.5
Butetown	33	20.1
Caerau	10	5.4
Canton	50	32.8
Cathays	22	13.4
Ely	9	6.1
Grangetown	22	13
Llanishen	31	19.5
Pontprennau	38	23.8
Rhiwbina	35	27.9
Riverside	32	20.1
Rumney	13	9.2
Splott	26	17
Whitchurch	37	26.8
Mean	27.57	17.83

Figure 10 Census data for education and occupation in selected wards of Cardiff.

Figure 11 Scatter graph representing the relationship between the bivariate data in Figure 10.

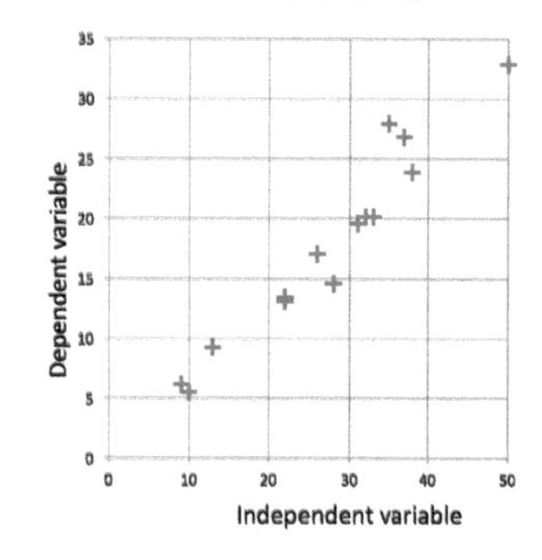

How do I tell if there is a correlation?

A scatter graph will show a correlation between the two variables if:

The dependent variable increases as the independent variable also increases. This pattern suggests that there is a **positive correlation** between the two variables. Figure 11 shows an example of a positive correlation.

The dependent variable decreases as the independent variable increases. This pattern suggests that there is a **negative correlation** between the two variables. Figure 12 shows an example of a negative correlation.

There is no correlation between the two variables if the points are scattered across the graph without making any pattern.

Figure 12 Scatter graph showing a negative correlation.

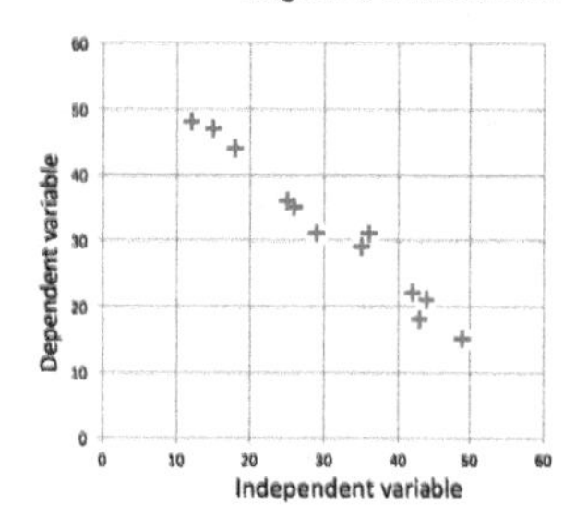

Learning objectives

- Understanding the difference between correlation and causality

Correlation and causality

Correlation implies that there is some connection between two sets of variables. However, the explanation for this connection may not be clear. For example, in studying the Census data for Cardiff, a student noticed that the percentage of households in inner urban wards who did not own a car was higher than in suburban wards. The student plotted a scatter graph with distance from the CBD on the horizontal axis. The graph showed a negative correlation. However, the explanation for this link is not clear. It could be that:

- Households in the inner city are poorer and cannot afford a car.
- Fewer homes in the inner city have driveways so there are fewer places to park a car.
- People living in the inner city have access to frequent buses and trains so don't need a car.

Causality means that there is a direct link between the two variables. This link can be used to explain why the value of the dependent variable changes. For example, in a study of 20 European cities, a student found a negative correlation between average July temperatures and latitude, that is distance from the Equator.

Activities

1 Study the list of paired variables in the table below. For each pair state:

a) which is the dependent variable and which is the independent variable;

b) whether you would expect to see a positive or negative correlation;

c) whether you think the relationship shows correlation or causality. Make sure you can justify each of your decisions

Fieldwork context	Variable One	Variable Two
A sandy beach	Wind speed (metres per second)	The height that sand is carried (by the wind) above the beach (in centimetres)
A transect up a hillside between 150m and 400m above sea level	Height above sea level (metres)	Wind speed (metres per second)
A pebble beach	The average size of sediment (millimetres)	Distance along the beach – measuring in the same direction as the movement of longshore drift (metres)
An eroded footpath in a honeypot site	Distance from the car park (metres)	Width of the area affected by erosion (metres)
Study of 10 different soil samples	Infiltration rate (cubic cms per minute)	Percentage of the soil composition made of clay particles
The area around a busy main road	Noise levels (decibels)	Distance from the road (metres)
A housing area close to an attractive urban park	Distance from the park (metres)	House prices (£1,000s)
A study of a river channel over several days	The discharge in the river channel (in cubic metres per second)	The amount of rainfall each day (in millimetres)
A larger urban area	Population density (number of people living per km^2)	Distance from the city centre (km)
A larger urban area	Population density (number of people living per km^2)	The percentage of households who do not own a car

Extrapolation and interpolation

You can use a line of best fit to estimate other values that could be expected if other points followed the same trend. This process is called:

- **Interpolation** if you estimate a value that is within the range of values plotted on your scatter graph. If most points on your graph are close to the line of best fit then interpolation should give a reliable estimate. However, the estimate will be less reliable if some points lie further away from the line of best fit.
- **Extrapolation** if you estimate a value that is outside the range of values plotted on your scatter graph. Extrapolation should give a reasonably accurate estimate if you try to estimate values that are only a little beyond the range of figures on your graph. However, you need to be cautious and consider whether your estimate is realistic. For example, on Figure 13, a ward with 2% of residents with a degree would have a minus figure for residents in professional occupations – this would be impossible!

Figure 13 How to interpolate and extrapolate

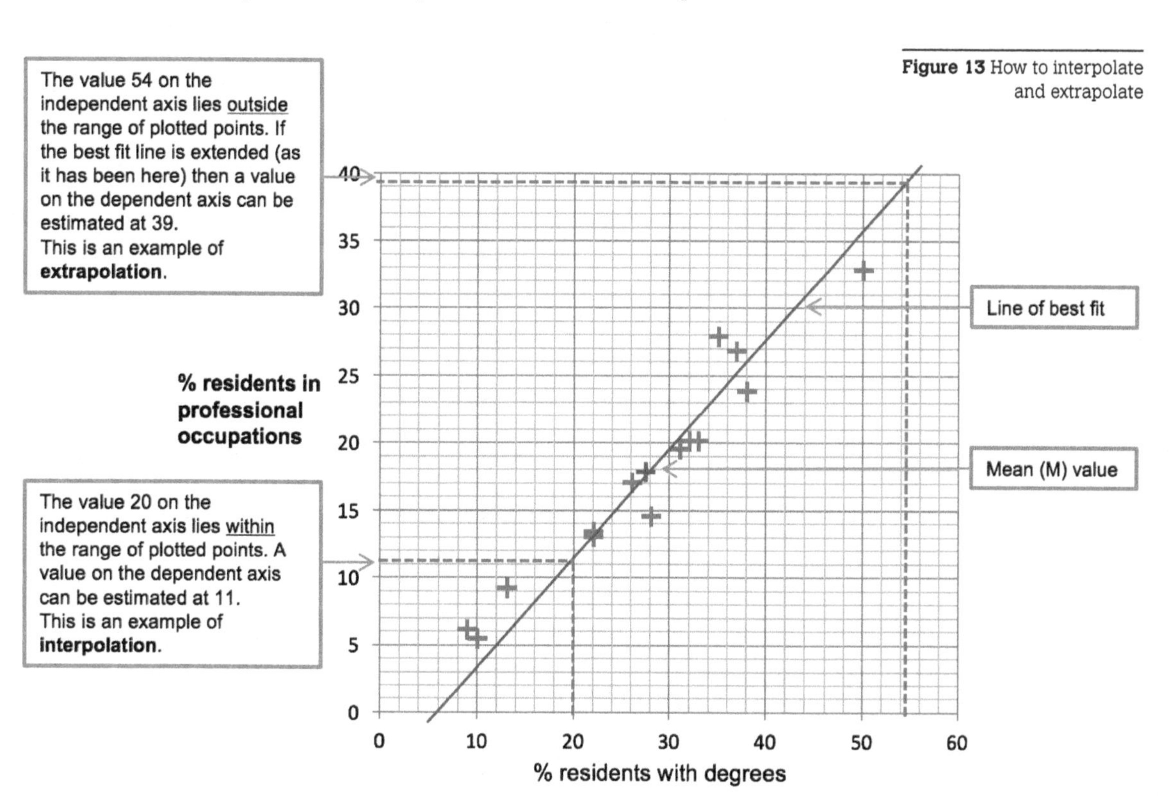

Chapter 5
Conclusion

Learning objectives
In this chapter we will explore:
- How to reach conclusions from your fieldwork enquiry.

How to reach a conclusion

You've collected data, processed, presented, and analysed it. It's time to end your enquiry by making your conclusions. A conclusion should make use of the evidence you have collected and analysed. To conclude an enquiry you need to:

1. remind yourself of the enquiry's aims;
2. consider the evidence;
3. draw together your overall findings for each enquiry question or hypothesis;
4. make a judgement/construct an answer to your original aim.

Step One Revisit the aims of the enquiry

It is important to remind yourself of the aims of your enquiry. The aims of an enquiry are often based on your understanding of a geographical issue, concept, or theory. What was the concept that you were investigating? What did you expect to find – did you make any predictions? In your conclusion you need to decide whether or not you have met your aims.

Step Two Consider the evidence

Think about each piece of evidence from primary and secondary sources. Each separate table, graph or map you have drawn is a piece of evidence and each tells a small part of the story. What patterns, trends, and connections can you see?

Step Three Draw together your overall findings

You will need to draw together all of the separate pieces of evidence from your enquiry. It is rather like making a jigsaw puzzle. You need to bring these pieces of evidence together so that you provide an overall conclusion.

Step Four Write your conclusion

When you write a conclusion you should refer to each piece of evidence. Remember to answer each hypothesis. Draw it all together by making a judgement, constructing an answer to your original aim.

Messy geography

The results of your fieldwork will not always be what you expected. When you look at the evidence, graphs may show no real trend and patterns on maps are not always neat and tidy. For example, you might think that a map of house prices for your local town would show that the lowest house prices are close to the town centre. You might also expect that your fieldwork would show that house prices get steadily more expensive as you move out towards the suburbs. If so, a map of house prices would show a regular (or systematic) pattern.
In reality, such a map is likely to be much more complex. It is likely to show clusters of higher house prices in inner urban areas – probably connected to positive features of the environment such as urban parks. In reality, the results of geography fieldwork are often messy rather than neat and tidy. Don't be surprised if your results don't exactly match your predictions.

Figure 1 A photograph of the study area. Borth, West Wales

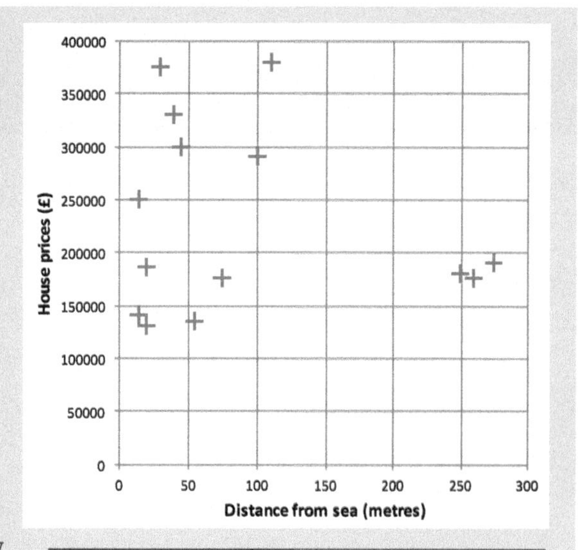

Figure 2 A scatter graph showing the possible relationship between distance from the sea and house prices in Borth

A group of students investigated the effect of new coastal defences at Borth, a small seaside town in West Wales. They investigated the concept of sphere of influence (see pages 96-101). They wanted to know:

- whether the coastal defences had positive or negative impacts on the town;
- how large the sphere of influence might be.

They conducted bipolar surveys to see what positive and negative impacts the construction of the new coastal defences might be. They also collected evidence of house prices to see whether the coastal defences created enough confidence for people to ask high prices for homes that were close to the sea. Their hypotheses were:

- *House prices are higher closest to the sea.*
- *Houses at sea level are more vulnerable to flooding (despite the defences) so are worth less than houses above sea level.*
- *The sphere of influence is affected by height above sea level rather than distance from the sea.*

Two key pieces of evidence are presented in Figures 2 and 3.

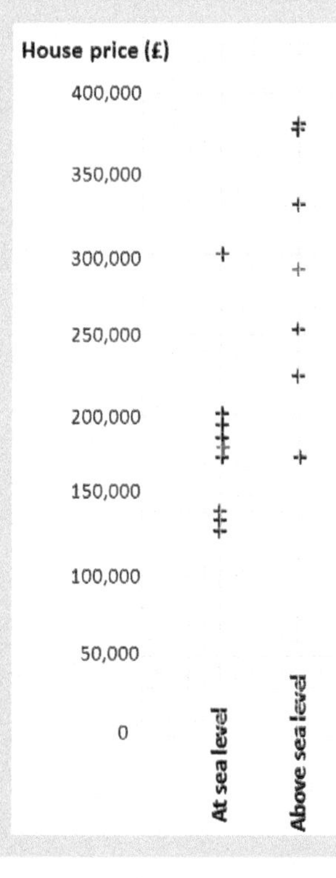

Figure 3 A dispersion graph showing the range of house prices at sea level (less than 3 metres above high tide) and above sea level (on the hillside in Figure 1)

Activities

1. **Describe the evidence shown in Figure 2.**
2. **Use Figure 3 to identify:**
   i) the median value of house prices in each location;
   ii) the range of house prices in each location.
3. **What conclusion can you draw when you put these two pieces of evidence together? Think about whether the students were able to answer any of their hypotheses.**

Chapter 6
Evaluation

Learning objectives
- Understanding how to evaluate the accuracy and reliability of your data collection.
- How to evaluate how well your enquiry process worked.

How to evaluate your enquiry

An evaluation should weigh up the fieldwork enquiry. Be methodical – think about each stage of the enquiry process and make a table of strengths and weaknesses.

Questions to ask yourself

When you begin to analyse your data it might be helpful to ask yourself some questions. For example:
- How far did I get?
- Am I any nearer to answering my enquiry question?
- Are there any other questions I should ask?
- Is there any reason to doubt the evidence?

Accuracy

Your choice of unit can improve **accuracy**. For example, if you are measuring height or distance it will be more accurate to measure to the nearest centimetre than the nearest metre. The smaller the unit, the more careful you need to be when reading from the apparatus. Inaccuracy may be due to errors when you read the apparatus or carelessness when the result has been written down.

More measurements will sometimes provide greater accuracy. For example, you will get a more accurate cross section if you take measurements at closer intervals along a transect.

Accuracy may also depend on how recently data was collected. Secondary data sources published on websites may be out of date. For example, data for the national census is only collected every 10 years.

Reliability

For data to be reliable it needs to be collected in such a way that, if the fieldwork is repeated, a similar set of data would be collected. **Reliability** is achieved by strictly following the same sampling procedures each time the data is collected.

Unreliable data has large variations when measurements are repeated. Wind speed and noise levels vary constantly so can provide unreliable data. You will get a more reliable measure if you take several readings and then calculate the mean.

Data will be unreliable if teams of students collect data using methods that are not consistent with one another. For example, unreliable data would be collected by teams of students if traffic was counted for 3 minutes at 10 locations but:
- the teams had not synchronised their watches so data was not collected at the same time;
- or some teams used stopwatches to keep to exactly 3 minutes but others did not.

Bias

Some data is affected by **bias**. Biased data may be factually incorrect, or its accuracy may be a matter of opinion. This is especially true of qualitative rather than quantitative data. If we analyse the evidence presented to us in the media (on TV, in newspapers, or in blogs) or in qualitative surveys such as questionnaires and interviews, we can identify that some data is presented as fact but is actually biased because it is someone's opinion. Identifying this bias is useful. It helps us describe the opinions of certain groups of people. By considering this bias we may also begin to understand why certain groups of people have strong opinions.

It is important to avoid writing questions that lead people to agree with your own point of view when designing a questionnaire (see pages 76-77). When you evaluate fieldwork that contains qualitative data collection you need to consider whether your method of data collection led to biased results.

Figure 1 How accurate and reliable are the results from this data collection?

These students measured river depth so that a cross section of the channel could be drawn. They measured down from a line that was stretched horizontally across the river channel.

They took depth readings at 1 metre intervals as they crossed the river. They used a rule to take the measurements and recorded the data to the nearest centimetre.

They didn't have any way of checking whether their line was actually horizontal. There was some slack in the line – you can see it in the photo. So, if they had repeated the task they would probably have found some significant variation in their measurements.

Dos and don'ts of evaluation

Do:

- ✓ Think about the strengths and weaknesses of your sampling strategy. With hindsight, did you choose the best strategy to give you data that represents the whole?
- ✓ Identify any possible inaccuracy or unreliability. Suggest how you might change the enquiry if you did it again.
- ✓ Identify bias and try to explain why certain groups of people provide biased evidence.

Don't:

- ✗ Simply suggest that you could have done more. A bigger sample might not have given you different results but different questions in your questionnaire might have been more effective.
- ✗ Write about what you enjoyed and didn't enjoy on the fieldtrip. That's not evaluation.

Activities

1 Study the information in Figure 1.

a) Describe one strength of this fieldwork.

b) Suggest one way this fieldwork could be made more accurate and one way it could be made more reliable.

SECTION 02

Chapter 7 Flows

Learning objectives

In this chapter, we will explore:

- how collecting data on geographical flows can help us to understand **physical** and **human** processes;
- how data on geographical flows can be recorded and presented.

What are geographical flows?

Geographical flows are patterns of movement. You can measure flows of:

- pedestrians;
- vehicles;
- wind speed;
- water in a river;
- longshore drift.

Why and how to measure flows of traffic and pedestrians

Traffic and pedestrian surveys can be used in **urban** or **rural** locations to provide evidence in an investigation about:

- how flows of shoppers change during the day;
- how the amount of traffic varies from one part of a town to another;
- the need for improvements to road safety or public transport provision.

Traffic and pedestrian counts will provide you with **discrete data** (see page 11). To conduct traffic counts, you will need to:

- design your data collection sheet;
- select the sites at which data will be collected;
- decide on the timing of the traffic counts;
- explain the technique to the other people who might be needed to conduct simultaneous counts. This is important so that your data is consistent and accurate. It will also allow you to map the data later.

How to record flows of traffic

The traffic count recording sheet usually has columns to record the different types of vehicles that pass by. It is common to use a tally chart (see pages 18-19) to count vehicles of different sizes. It is also usual to record the direction of travel, for example, into the city centre or out of the city centre.

Figure 1 An example of a vehicle count sheet

Time: From: _____ To: ___________ Location: ___________________

Weather: __________

Bicycles		Motorbikes & scooters		Cars		Vans		Lorries	
In	Out	In	Out	In	Out	In	Out	In	Out

Strengths and limitations of measuring flows of pedestrians and traffic	
Some strengths	**Some limitations**
Working as a group it is possible to conduct simultaneous surveys at a number of locations so that spatial patterns can be identified and analysed. Counts at different times of day can be used to analyse temporal patterns (change over time). Additional information, such as vehicle occupancy, or the home-base of lorries (based on information written on the vehicles), can also be collected.	Pedestrians can be difficult to count accurately because their movements are unpredictable. Counts done by different people will give inaccurate results if they count at slightly different times, or for different lengths of time.

Recording flows of pedestrians

You might want to count flows of pedestrians in different shopping streets. This might help an enquiry that is investigating the quality of the retail environment and why shoppers prefer some streets. You might also want to count flows of visitors at a honeypot site, like the upland area shown in Figure 8 on page 33.

Pedestrians can be more difficult to count than vehicles as their movement is more unpredictable. For example, they may change direction or walk in groups with few gaps between them. This can make it difficult to count them accurately. So, quantitative data can be collected but you may need a strategy to improve the accuracy of your count, especially in busy pedestrianised shopping areas. One strategy is for three people to count at the same location for five minutes. Then, calculate the average (see pages 28-29).

As with traffic counts, tally charts are usually used to record pedestrians. A decision would have to be made whether to record total number of pedestrians, or to record groups of pedestrians based on, for example, their age or gender. You would only record the age or gender of pedestrians if it was relevant to your enquiry question or hypothesis.

Figure 2 Pedestrianised street in Cardiff. The results of a pedestrian count in this environment could be unreliable because of the number of people and the complexity of their movement

Activities

1 **Study Figure 2. Students wanted to see whether family groups stayed in this square for longer periods of time than single people.**

a) Design a sampling strategy that could be used to count pedestrians in this street. Decide how many people you would need and where they might stand.

b) Design a data collection sheet for this location. It must allow the user to record evidence that will help.

Measuring flows of water

We might want to investigate the speed (**velocity**) of a river to find out if:

- flows are faster on the inside or outside of a **meander**;
- velocity increases or decreases downstream;
- velocity of flow is affected by human management of the river course.

Measuring the velocity of a river's flow is an example of collecting **continuous data** (see page 11).

The simplest and quickest method is to use a float. You will need:

- a float, for example, a dog biscuit;
- a tape measure;
- a stop watch;
- Wellington boots and waterproof clothing.

Step One At each data site, one person carefully places the float in the river. A second person, standing 10 metres downstream, records the time taken for the float to reach them.

Step Two Record the time it takes the float to travel 10 metres downstream at 10 equally spaced intervals across the river.

Step Three Calculate an average time, by adding the results together and dividing the result by 10.

Step Four Calculate the velocity by dividing distance by average time, for example, if the average time was 48 seconds then the velocity would be 10 metres divided by 48 seconds, giving a stream velocity of 0.208 metres per second.

Figure 3 A student measuring river flow with a flow meter

A more precise method of measuring velocity is to use a flow meter. A float moves across the surface of the water so it is only the surface velocity that is being measured. A flow meter is more accurate because it is placed below the surface of the water. The flow meter should be put in the water for a set time, for example, 1 minute and, as with the float method, placed at regular intervals, usually 10 locations, across the channel.

The operator must stand either beside or downstream from the flow meter. If they stand upstream of the flow meter, they will interfere with the river's natural flow and the reading will be unreliable.

Strengths and limitations of measuring river velocity	
Some strengths	**Some limitations**
The method is straightforward to do. Taking 10 measurements across a channel and calculating the mean increases reliability of results. It provides data that could be analysed to find connections with other variables such as river depth, rainfall, or pebble size.	Using a surface float means that only the surface velocity of the water is being measured and recorded. Flow meters are fragile and expensive pieces of equipment. It is possible to disrupt normal water flow during measurement and, therefore, collect data that is inaccurate.

Dos and don'ts of measuring flows in a river or on a beach.	
Do:	**Don't:**
 For any fieldwork on beaches and near rivers, safety is critical. It is important to work in groups, to have an adult supervising, and to find out relevant information about tide times and weather forecasts beforehand.	✗ Get in a river unless you are supervised. Never get in a river above knee height.

Why and how to measure infiltration

Infiltration is the downward movement of water into the soil. **Infiltration rate** is the speed at which water enters the soil. It is usually measured by the depth (in mm) of water that can enter the soil in one hour. Measuring infiltration rates would be useful in investigations into the water cycle. For example:

- why some areas are more prone to flooding than others;
- how rainfall, infiltration rates, and river velocity are connected.

Figure 4 How to measure infiltration – the flow of water into the soil

To measure infiltration a cylindrical piece of equipment called an infiltrometer should be used. An infiltrometer can be made from a piece of drainpipe. This is very straightforward and can be fun! You will need:

- an infiltrometer, or a piece of drainpipe at least 30cm long;
- a ruler;
- a water container;
- a stopwatch.

Step One Knock the infiltrometer/drainpipe into the ground with a piece of wood and hammer. Place the ruler inside the drainpipe so that it touches the ground surface.

Step Two Person 1 fills the drainpipe with water.

Step Three Person 2 starts the stopwatch and records how long it takes for the water to drop each centimetre inside the drainpipe.

Step Four Repeat the experiment a number of times in different places to see how much the infiltration rate varies in the local area.

Strengths and limitations of measuring infiltration	
Some strengths	**Some limitations**
The equipment required is low cost or can be homemade. Comparison of infiltration rates on different types of natural and urban surfaces can be made. Refilling the drainpipe and repeating the experiment can show how the speed of infiltration is affected as the soil becomes saturated. The experiment can be repeated after heavy rain to see how the infiltration rate is affected by weather conditions.	Factors such as previous rainfall or current weather, for example, temperature and wind speed, can greatly affect results. Water pressure will drop as water infiltrates, thus affecting speed of infiltration. This can be taken into account by topping up the water, keeping a note of how much is added. Care needs to be taken to avoid damage to vegetation and other land surfaces.

Activities

1 Design a data collection sheet for the infiltration rate investigation described on this page.

How to collect data about flows in a coastal environment

Longshore drift is the movement of sediment (such as sand or pebbles) along the coast. By collecting information, such as depth of sediment and movement of materials in the sea, we can assess whether longshore drift is happening and, if so, in what direction and at what speed.

Groynes trap sediment to prevent it being moved further along the coast by longshore drift. Measuring the build-up of sediment on either side of the groynes should provide evidence of the direction of longshore drift.

Figure 5 A survey of sediment on a beach

Step One Measure each groyne. Collect data at three points along the groyne at 0m, 20m, and 40m.

Step Two Measure the distance from the top of the groyne to the level of sediment on each side of the groyne at each of the three sample points. Record this data on the data recording sheet.

Step Three Take measurements at the same sample point along each groyne.

Step Four Process the data by calculating the mean distance (from the groyne to the beach) for each groyne by adding the three results together and dividing by three.

Figure 6 Data collected on a beach in NE England

Groyne	North side Mean distance to sediment (cm)	South side Mean distance to sediment (cm)	Difference in mean height (cm)
1	150	100	
2	120	80	
3	120	78	
4	140	88	
5	138	70	
6	132	56	
7	110	92	
8	56	54	
9	68	60	
10	82	72	

How to measure speed of transport of material in the sea

A float in the sea can provide an indication of the speed and direction of longshore drift. Equipment needed to measure the speed and direction of movement:

- a float such as an orange, or a painted tennis ball;
- a stopwatch (you could use the stopwatch app on your phone);
- a tape measure;
- a data collection sheet

Step One Place the tape measure as close to the sea as is safe, parallel to the shore line.

Step Two Mark a set distance, for example, 10 metres, perhaps with poles.

Step Three Place the float gently into the sea adjacent to the start point.

Step Four Measure and record the time it takes the float to travel 10 metres.

Step Five Repeat this process a number of times and calculate an average time.

Figure 7 How to measure the speed and direction of longshore drift

Strengths and limitations of measuring speed and direction of longshore drift	
Some strengths	**Some limitations**
The equipment required is low cost. An orange as a float is biodegradable (but as it would take a long time to decompose, it is better to try and retrieve the orange). The speed of movement of the float may be affected by wind speed and direction so these variables should also be recorded. Results for areas of managed and unmanaged coastline can be compared.	Rocky outcrops in the water may obstruct/deflect the float. Swimmers or people paddling in the water may affect the movement of the float. If the float is thrown into the water, it will gain extra momentum and your result will be unreliable. The float may be dragged out to sea by the backwash and get lost.

Activities

1 **Use Figure 6.**
   a) Calculate the difference in sediment height either side of the groynes.
   b) Decide and justify whether it would be better to present these results using line graphs or bar charts.
   c) Explain what the results suggest about the direction of longshore drift on this stretch of coastline.

2 **Discuss how the limitations listed above could be reduced.**

Time to practise questions about flows

1 Students decided to collect evidence about the movement of pebbles on a beach in NE England. They placed 40 pebbles (numbered and brightly painted with waterproof paint) in an area of the beach where there were few obstacles to the movement of the pebbles. The start location of the pebbles was marked with poles. The students returned the next day and found 12 of the pebbles.

Figure 8 Results of pebble movement on a beach

Pebble No.	Weight (grams)	Distance along beach N or S (metres)
1	180	0.8 N
2	326	11.6 S
3	356	3.4 N
4	322	14.1 S
5	408	5.6 N
6	206	13.1 S
7	294	16.3 S
8	418	2.9 S
9	506	0.5 S
10	318	12.2 S
11	336	4.2 S
12	524	0.6 S
		Mean distance moved = 7.1m

Study Figure 8.

a) Suggest a graphical technique that could be used to assess the relationship between the weight of the pebbles and the distance they had moved. Explain why you think this would be a suitable technique. [4]

b) Study the results of pebble movement along the beach in Figure 8. What can you conclude about the process of longshore drift on this beach? [4]

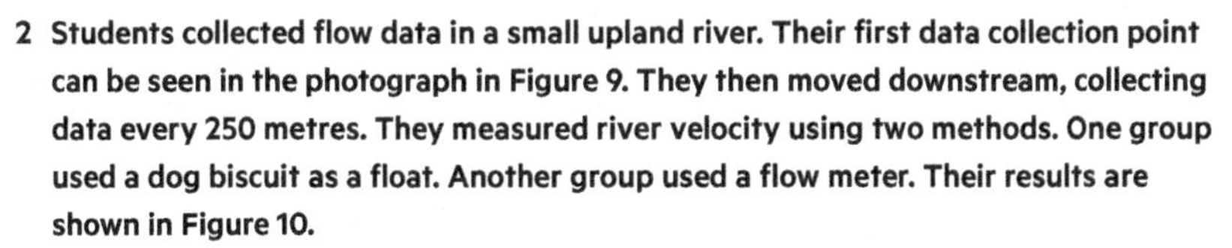

2 Students collected flow data in a small upland river. Their first data collection point can be seen in the photograph in Figure 9. They then moved downstream, collecting data every 250 metres. They measured river velocity using two methods. One group used a dog biscuit as a float. Another group used a flow meter. Their results are shown in Figure 10.

a) Suggest one advantage and one disadvantage of each of these two methods of measuring river velocity. [4]

Figure 9 Cardingmill Valley, Shropshire

Site Number (from upstream to downstream)	Velocity (m/second) Measured using dog biscuit	Velocity (m/second) Measured using flow meter
1	0.44	0.08
2	0.41	0.34
3	0.41	0.40
4	0.51	0.34
5	0.66	0.14
6	0.70	0.34
7	0.66	0.50
8	0.33	0.39
9	0.44	0.60
10	0.33	0.39
11	0.84	0.81
12	0.86	0.73
13	0.91	0.76
14	0.73	0.58
15	0.83	0.85

Figure 10 Results of velocity measurements on a small river in Shropshire

b) Suggest how the data in Figure 10 could be presented. Justify your choice. [4]

c) What do the results in Figure 10 suggest about the relationship between distance downstream and river velocity? [2]

3 A group of students counted the number of cars each morning and evening at a busy urban road junction where 4 roads go north, east, south, and west from the town centre. The students did the counts for *5 minutes* each time. Data was collected for 7 days and the average number of cars passing each location was calculated. The results are shown in Figure 11.

Figure 11 Traffic Count near a busy town centre

Road going North from town centre

	Morning (in)	Morning (out)	Evening (in)	Evening (out)	TOTAL
Average number of cars	52	33	30	50	165

Road going East from town centre

	Morning (in)	Morning (out)	Evening (in)	Evening (out)	TOTAL
Average number of cars	31	21	30	35	117

Road going South from town centre

	Morning (in)	Morning (out)	Evening (in)	Evening (out)	TOTAL
Average number of cars	43	22	20	50	135

Road going West from town centre

	Morning (in)	Morning (out)	Evening (in)	Evening (out)	TOTAL
Average number of cars	57	34	30	55	176

Study Figure 11.

a) Construct a simple flow line map to show the number of cars *per minute* entering and leaving of the town centre in the evening. Use a scale of 1mm = 2 cars. [6]

b) Calculate the *total* number of cars *per hour* entering the town centre in the morning. [2]

Chapter 8
Transects

Learning objectives

In this chapter we will explore:

- methods of doing long transects and belt transects;
- the use of kite diagrams to present transect data;
- the use of transects in urban areas.

Why use transects?

To use a **transect** means to sample along a line. Transects are useful in both physical and human geography if you want to investigate how something varies over distance. For example, you could use transects to investigate each of the following hypotheses.

- Pebbles get smaller as you go along a beach.
- Wind gets stronger as you go up a slope.
- Quality of life improves as you get further away from the city centre.

You can use any of the four main sampling strategies to collect data along each transect (see pages 12-13). Before sampling along any transect you will need to take the following steps.

Step One Do initial research to identify the most appropriate transect sites and decide where the transect should begin and end.

Step Two Decide on the sampling strategy which best meets your needs. This will depend on the length of your transect and the variable you are measuring.

Step Three Select equipment to collect the data. For example, you may need an app to record elevation or a noise meter. You may want a digital camera or mobile phone so that you can collect photographic evidence.

Step Four Design data collection sheets.

Why and how to do a belt transect

Belt transects are used to sample data across small features. They usually involve the use of quadrats (see pages 14-15). Quadrats are metal frames, often half a metre square, which are divided into small squares. The number of variables (for example, species of plant) within the area of the quadrat can be counted. You should take control readings before beginning the belt transect. This provides a control against which you can compare the data collected along the transect.

Figure 1 Continuous belt transect (top) and interrupted belt transect (bottom)

Data collection along belt transects can either be continuous or interrupted. A continuous belt transect is made by turning each quadrat over, end-to-end, along the transect An interrupted belt transect is made by placing the quadrats at regular intervals along the transect. Figure 1 shows each type. Interrupted belt transects are used when the feature being studied covers a relatively long distance.

Strengths and limitations of using quadrats in a belt transect	
Some strengths	**Some limitations**
Belt transects enable detailed information about the amount, as well as presence, of variables to be collected at each sampling location. Quadrats are divided into smaller squares so that it is possible to estimate the percentage distribution of whatever feature is being measured.	Interrupted belt transects can save time but important variations in the data may be missed in the gaps between samples.

Distance across footpath (cm)	C1	0	50	100	150	200	250	300	350	400	450	500	550	600	C2
% bilberry	75	60	30	0	0	0	0	0	0	0	0	10	35	60	55
% heather	15	25	15	0	0	0	0	0	0	0	0	0	10	25	30
% grass	0	0	10	10	5	0	10	25	20	0	0	5	5	0	5
% other vegetation	10	15	15	10	5	0	5	5	0	0	0	10	5	15	10
% bare ground	0	0	30	80	90	100	85	70	80	100	100	75	45	0	0

C1 and C2 are control quadrats sampled at locations away from the footpath.

Figure 2 Results of a continuous belt transect across a footpath in a honeypot location in Shropshire Hills AONB

How to draw a kite diagram

One suitable technique that can be used to present information of percentage distribution of variables collected using quadrats is a **kite diagram**. To plot a kite diagram, take the following steps.

Step One Draw a series of horizontal lines across a piece of graph paper. Each horizontal line corresponds to the centre line for each one of the species for which data was collected. The length of the horizontal lines corresponds to the total length of the transect. The vertical scale corresponds to the percentage of each species present.

Step Two For each species, plot points equidistant from the horizontal central line, equivalent to half of the value recorded on the data collection sheet. Each point is plotted along the line at the distance where the data was collected.

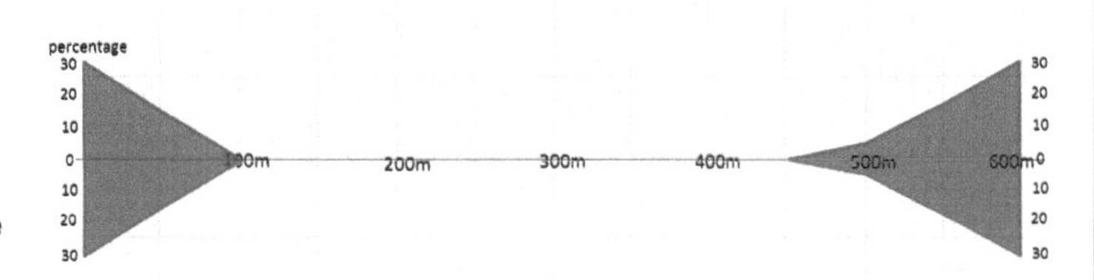

Figure 3 A kite diagram showing the percentage of bilberry along the transect in Figure 2

Step Three Join all the points above the horizontal centre line with a ruler. Do the same for the points below the horizontal centre line. Colour the shape between the lines you have drawn.

Step Four Repeat this process for each species. You should use different colours for each species.

Activities

1 **Study Figure 1. Suggest one advantage and one disadvantage of using the continuous belt transect to investigate how plants are affected by trampling by walkers.**

2 a) Draw kite diagrams to show the percentage of heather, grass, other vegetation and bare soil across the footpath.

b) What conclusion can you reach about the effect of trampling on these plants?

Using transects to draw cross sections

Transects can be used to create a **cross section** across a feature. A cross section represents the shape, or **profile**, of a feature. To create one, you need to measure distances and depths from a horizontal line. Alternatively, you may be able to use a clinometer, which measures angles, to record the gradients along your cross section. Pages 16-17 show how to use a clinometer to create a **beach profile**.

You may decide to sample other variables, such as wind speed, noise, or type of vegetation, along your transect. This data would enable you to do two things:

1 draw a cross section to show the profile of your feature;
2 plot extra data onto the cross section about other variables.

Figure 4 Sand dunes at Ynyslas, West Wales

For example, study the environment in Figure 4. You could set up a transect through the dunes, starting at the top of the beach and working away from the sea. You could use a clinometer to measure gradients and construct a cross section. You could also collect wind speed data using an anemometer at the top of each dune and in between each dune. This would allow you to reach conclusions about how wind speeds are affected by height of the dunes *and* by distance from the sea.

How to investigate the cross section of a river channel

You might want to use a transect to investigate how data varies across the width of a river channel. For example, a transect would be useful in either of these enquiries.

- *Is the water flowing fastest on the outside of the bend?*
- *Are pebbles sorted by deposition so they get smaller up the slip of slope?*

To create a cross section of the river channel you will need to take the following steps.

Step One Measure the width of the river channel. Divide this by 10 to create 10 evenly spaced sample points. In Figure 5 the channel is 3m across, so the samples are 30cm apart.

Step Two Stretch a line tightly across the river. Make sure it is level. Include some of the river bank on either side so that you can draw the river's banks on your cross section.

Step Three Measure downwards, from the line to the ground, at each sample point. In Figure 5, measurements were made every 30cm on each side of the river as well as into the river channel. Record the distance from the line to the surface of the water.

Figure 5 How to take data readings to create a cross section across a river channel. Shropshire Hills AONB

How to investigate soil erosion across a footpath

Cross sections can be used to investigate the impact of trampling at a **honeypot site**. These are places that receive a lot of visitors, as in Figure 6. You can use the method described in Figure 5. Remember that to show footpath depth you will need to plot the depth figures downwards along the y axis. You could combine this technique with the belt transect described on page 64.

Figure 6 How to take data readings to create a cross section across an eroded footpath. Shropshire Hills AONB

Distance along transect (cm)	**0**	**100**	**200**	**300**	**400**	**500**	**600**	**700**	**800**	**900**
Distance from line to ground (cm)	25	27	29	45	52	47	42	40	26	24

Activities

1 **Suggest why it is important that the line used in Figures 5 and 6 must be tight and level.**

2. **Use the data in Figure 6 to draw a cross section of the footpath.**

Longer transects or long profiles

To investigate change over longer distances you will need a longer transect. For example, you might want to investigate how a river changes along its course. You might collect data on several variables such as river width, water velocity, and gradient. A sample along a lengthy feature like this is sometimes described as a **long profile**.

Longer transects are often used to investigate variations in environmental quality across urban areas. This may involve the use of an Environmental Quality Index (EQI) (see pages 20-21). For example, you might use a transect to collect data at increasing distance away from an urban feature such as a park, busy road, or building site to investigate the impacts of that feature on the locality (a **sphere of influence** study). Alternatively, you could create long profiles of two or more routes across a town to investigate sustainable traffic management, or the safety of pedestrians and cyclists.

http://www.sustrans.org.uk/

Fieldwork assessing cycle routes to school - using mobile apps

A very popular fieldwork enquiry is to investigate how your school promotes a more sustainable policy for students and staff to travel to school. Follow the weblink for more detailed background information.

Most schools have travel policies to promote sustainability. It is estimated that, across the UK, families could save £520 million per year on car reliance if students cycled to school instead of arriving by car. Other benefits are lower exhaust emissions, less congestion around schools, reduced fuel consumption, and a healthier student population. A slightly amended version of this enquiry could also be used to investigate how risk can be reduced in an urban environment by choosing the safest cycle route to school.

Surveying sustainable cycling routes

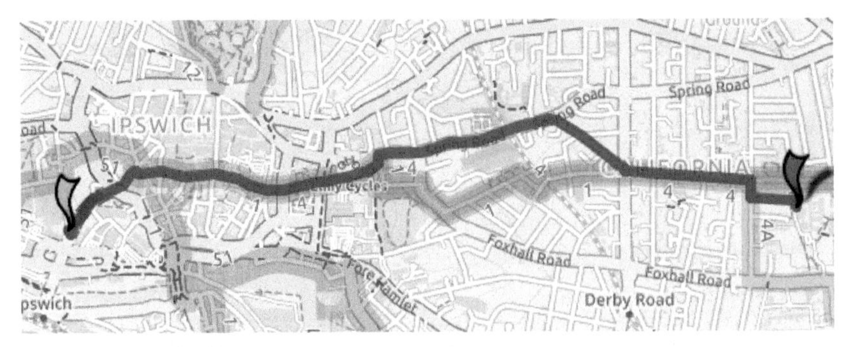

Figure 7 Screenshot from the Hub Bike mobile app

There are several free mobile apps available to generate suggested cycling routes. The screenshot shows the fastest cycling routes from home to secondary school. The app allows you to choose alternative routes, such as the quietest route, or the most balanced route in terms of gradient. You can use the interactive maps within the app to see sections of the map in great detail. The PC version will also provide you with a long profile to show gradients along the route.

Step One Download the free app to your mobile. Search for 'Hub Bike' in this case.

Step Two Plan the route by inputting your home postcode and your school postcode.

Step Three Select the type of route you want to survey.

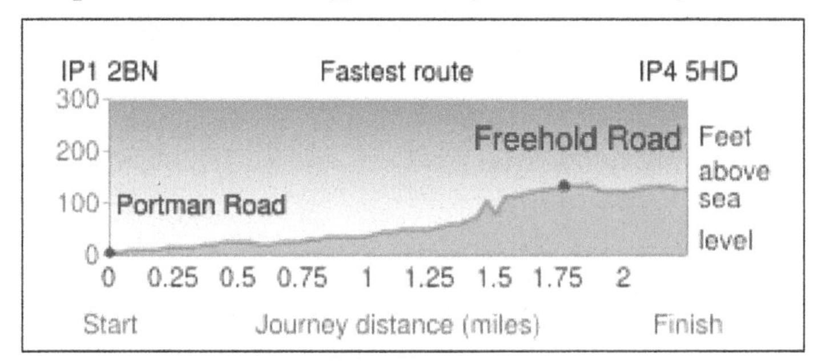

Figure 8 The long profile of the route

For the same start and end points as Figure 8, Figure 9 shows the quietest route for cyclists to use. The long profile was created automatically when the route was selected in the app. Figure 9 shows annotations added by a student after they walked the route on a fieldtrip. Important issues affecting cyclists are highlighted. The technique of adding annotations can be applied to all types of field investigations that make use of cross sections.

Figure 9 The recommended quietest route for the selected cycle journey.

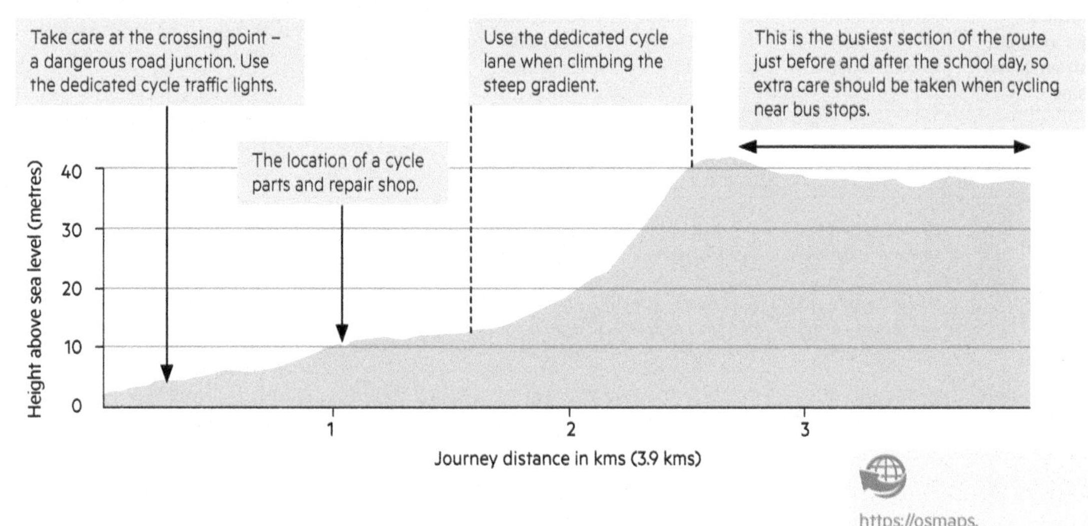

https://osmaps.ordnancesurvey.co.uk/
Follow the guidance for creating a route.

More detailed long profiles which show gradients due to change in altitude can be created by accessing the Ordnance Survey website

Strengths and limitations of long profiles	
Some strengths	**Some limitations**
Long profiles don't necessarily follow straight lines so are particularly useful where it is not possible to sample in a straight line due to street layout, or where access is prevented by private land-use. Long profiles are useful for showing changes in gradient (for example, along a river).	Long profiles record changes over longer distances than usual transects so are time consuming and you may need transport.

Time to practise questions about transects

Criteria Feature	5	4	3	2	1
Vegetation	Plenty of trees. 1+ tree per 20m	Some trees. 1 tree per 20-40m	Few trees. 1 tree per 40-80m	Sporadic trees. 1 tree per 80-100m	No trees or greenery
Traffic flows	No traffic	Some traffic but free flowing	Much traffic but free flowing	Much traffic and some congestion	Much traffic congestion. Stationary traffic blocking roads
Parked cars	Most properties have garage or parking space. Few parked cars on road	Some parking provision. <3 cars per 100m	3-6 cars parked per 100m	7–10 parked cars per 100m	Poor parking provision. >10 parked cars per 100m
Litter	No litter	Some litter outside <10% of buildings	Some litter outside 10-25% of buildings	Much litter. Litter outside 25-50% of buildings	Litter outside >50% of buildings
Private open space - gardens	All houses have private open space	About 75% have private open space	About 50% have private open space	About 25% have private open space	No private open space. Doors open onto street
Street furniture, for example, street lights, bins	All street furniture well-designed and in good condition	About 75% well-designed and in good condition	About 50% well-designed and in good condition	>75% badly designed or vandalised	All badly designed or vandalised
Pavements	All pavements well-maintained	Pavements in good condition but narrow	Some pavements damaged <25%	Significant pavement damage: 25-50%	>50% pavements in poor condition
Signage	Signage well-placed and well-maintained	Well-placed but not well-maintained	Some signage that is well-placed and maintained	Limited signage in poor condition	Inadequate poor quality signage or no signage
Building quality	All buildings well-designed and well-maintained	>75% buildings well designed and well-maintained	About 50% of buildings well-designed and well-maintained	>75% of buildings in poor condition.	All of buildings poorly designed and in poor condition
Noise	Very quiet – no significant noise	Occasional noise	Some noise at certain times	Significant levels of noise at times	Significant continuous noise intrusion

Figure 10 An environmental quality index (EQI) sheet designed by students who were investigating whether the quality of residential areas changed from the centre of the town towards the suburbs.

1 Study Figure 10.

a) Design a set of criteria for one more feature that could be added to the EQI to evaluate access to retail services. [4]

b) Explain why scores given for the environmental quality features in Figure 10 could be affected by the time of day, week or year. [6]

c) Explain why there may be limitations in conducting only one EQI transect survey from the CBD to the southern edge of the town. [4]

2 **Students conducted belt sampling through the sand dune shown in Figure 4 on page 66. They decided that rather than using systematic sampling, they would collect information on each dune ridge and in each dip. The students used quadrats to record the number of species at each site and collated their results in two tables, Figure 12 and Figure 13.**

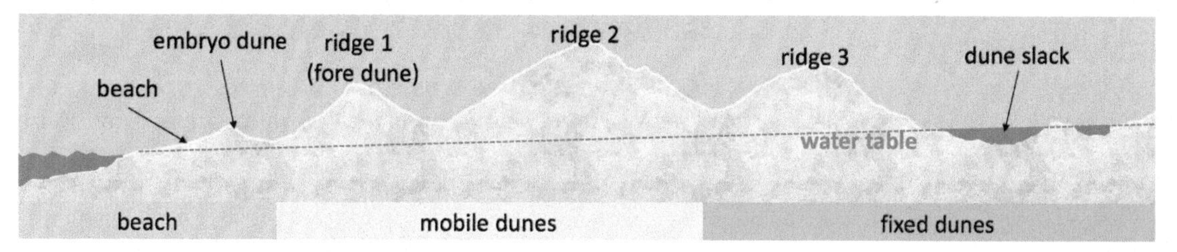

Figure 11 Plants vary with distance from the sea creating zones within the sand dunes

Plant species	Distance from sea (m)													
	0	20	40	60	80	100	120	140	160	180	200	220	240	260
Sea couch grass		20	10											
Marram grass			60	80	75	60	10	10	10					
Red fescue				10	15	10	10	20	15					
Bell heather						20	30	10	25	10				
Creeping willow											20	35	45	20
Other plants						10	35	40	35	80	80	65	55	80
No plants – only sand	100	80	30	10	10		15	20	15	10				

Figure 12 Percentage of plant species recorded at each sampling location

Plant species	Mobile dune	Fixed dune
Sea couch grass	8	0
Marram grass	20	8
Red fescue	12	20
Bell heather	16	28
Creeping willow	0	16
Total number of plants	56	72

Figure 13 Total number of plants sampled on mobile and fixed dunes.

a) Describe how the information in Figure 12 could be represented graphically. [4]

b) Justify your choice of graphical technique. [4]

c) The information in Figure 13 could be presented as two pie charts. Describe how the pie charts could be modified to show the difference in total number of species on mobile and fixed sand dunes. [4]

d) What conclusions can you reach about the differences in plant distribution on the mobile and fixed sand dunes? [6]

Chapter 9
Qualitative Data

Learning objectives

In this chapter we will explore:

- three methods of collecting qualitative data - **bipolar surveys, Likert surveys, and questionnaires;**
- ways in which qualitative data can be recorded and presented.

What is qualitative data?

Qualitative data is anything that helps us to describe a place, or geographical issue, with words or pictures rather than by measuring with numbers. We collect qualitative data when we want to know:

- how people think about (or perceive) a place;
- whether they agree or disagree with a point of view;
- someone's attitude towards an issue, for example, whether a new shopping centre should be built.

We can collect qualitative data by taking or collecting photographs and other images, interviewing people, or asking people to fill out a form or questionnaire.

Qualitative data usually starts with words or pictures but words or types of response can be counted. This makes qualitative data easier to record, present, and analyse.

Key things to decide when collecting qualitative information

- What information is needed to investigate the hypothesis or enquiry question.
- Whether you need to collect evidence from different groups of people, for example, residents, visitors, people of different ages.
- How many people should be surveyed in order to have a useful set of results.
- Whether the survey needs to be conducted at a number of different locations.

Why use a bipolar survey?

You can use a **bipolar survey** to gather information about how people think about a place. Bipolar surveys (and Likert surveys) provide respondents with statements from which they choose their response. This technique is called **closed** questioning.

Step One Write a number of pairs of statements which are 'opposite' to each other.

Step Two Design your data collection sheet. Examples are shown in Figures 1 and 2. The positive numbers represent favourable attitudes towards the place. The negative numbers represent negative attitudes towards the place. Using numbers in this way will allow you to quantify your results which will make it easier to analyse the evidence.

Step Three Some statements in your bipolar survey may seem more important than others. Study Figure 1. The ability of a sea wall to prevent flooding seems more important than how attractive it is. If so, this limitation can be reduced by weighting (increasing) the score values for the feature(s) considered to be more important (see page 21 for more on weightings).

Step Four Decide on your sampling strategy and sample size. You could use the bipolar survey to find the views of:

- your classmates;
- members of the public.

Positive Attitude to Feature	+3	+2	+1	-1	-2	-3	Negative Attitude to Feature
Access to the beach is not affected							Disrupts access to the beach
Likely to stop/ reduce flooding							Unlikely to stop/ reduce flooding
Protects or enhances the habitats of plants and animals							Disturbs the habitats of plants or animals
Low cost of building and maintenance							Very expensive to build and maintain

Figure 1 This bipolar survey could be used to test people's perception of a coastal defence scheme. Borth, West Wales

Positive Attitude to Feature	+2	+1	0	-1	-2	Negative Attitude to Feature
Housing well-maintained						Housing in poor condition
Little traffic - not a hazard to pedestrians						Too much traffic
Pavements well-maintained; easy to move around						Pavements uneven and/or congested
Area feels safe and secure						Area feels unsafe

Figure 2 This bipolar survey could be used to test people's perception of a residential area. Grangetown, Cardiff

Activities

1. **Study Figures 1 and 2. Add two more pairs of bipolar statements that could be used in each of these fieldwork locations.**
2. **Study the two bipolar surveys above. One uses zero in the middle of the range of numbers: the other does not.**
   a) Discuss how using a zero might affect your results.
   b) Decide whether it is better to use a zero or not. Justify your choice.

Likert Surveys

Likert surveys (pronounced lick it) are another way of finding out how people feel about a place or geographical issue. Respondents are asked whether they agree or disagree with a range of statements.

Step One Create your statements. For example, if the investigation is about the impact of tourism on a honeypot location, the following statements could be put to respondents:

- Tourism creates a significant number of jobs in the local area.
- Local farmers benefit from the presence of tourists.
- Tourists do not have an adverse impact on the natural environment.
- Tourism does not cause adverse traffic problems in the area.
- Tourism encourages the purchase of second homes.

Step Two Create a data collection sheet so you can record whether respondents:

Strongly agree, Agree, Are Neutral, Disagree, or Strongly Disagree with each statement.

You can give numerical values to each choice, as shown in shown in Figure 3. This will allow you to analyse your results more easily.

Figure 3 Values for a Likert survey

Strongly agree	Agree	Neutral	Disagree	Strongly disagree
5	4	3	2	1

How can information from Bipolar and Likert surveys be presented?

Data collected in bipolar surveys and Likert surveys can be presented in a wide range of graphical forms. Imagine 10 people were asked to complete bipolar surveys (as shown in Figure 1 on page 73) to compare attitudes towards sea walls and groynes. A set of results is shown in Figure 4 below.

Figure 4 Results for a bipolar survey
Key S = seawall and G = groyne

Positive Attitude to Feature	+3	+2	+1	-1	-2	-3	Negative Attitude to Feature
Attractive: enhances the environment	G	GG	SSS GGG	SS GG	SS GG	SSS	Ugly: makes the area unattractive
Access to the beach is not affected	GGGG	SS GG	SS GG	SS G	SS G	SS	Disrupts access to the beach
Likely to stop/reduce flooding	SSSS	SSS	SS GG	S GG	GGG	GGG	Unlikely to stop/reduce flooding
Likely to stop/reduce erosion	SSS GGGG	SSS GGG	SS GG	S G	S		Unlikely to stop/reduce erosion
Protects or enhances habitats of plants and animals	SS GG	S G	S GG	SS GG	SS GG	SS G	Disturbs habitats of plants or animals
Low cost of building and maintenance	S GGG	S GG	S GG	S GG	SSSS G	SS	Very expensive to build and maintain

How to present information from a Likert survey

A dispersion graph can be used to show the patterns of distribution of different sets of data. Where a comparison is being made of two sets of data (as in Figure 4) the points can be plotted, as in Figure 5, using different symbols, on one graph. Alternatively you could plot two separate graphs.

To present the information in Figure 4 take the following steps.

Step One Draw an x axis and y axis on the graph paper.

Step Two Divide the x axis into the number of columns needed to plot each feature for which data has been collected (in Figure 4 there are 6 features used to assess opinions so 6 columns are needed).

Step Three Along the y axis, plot the possible scores (in Figure 4 the values go from +3 to -3).

Step Four Plot each result in the appropriate column. Results of the same value are plotted side by side.

Figure 5 A dispersion diagram showing the results of Figure 4

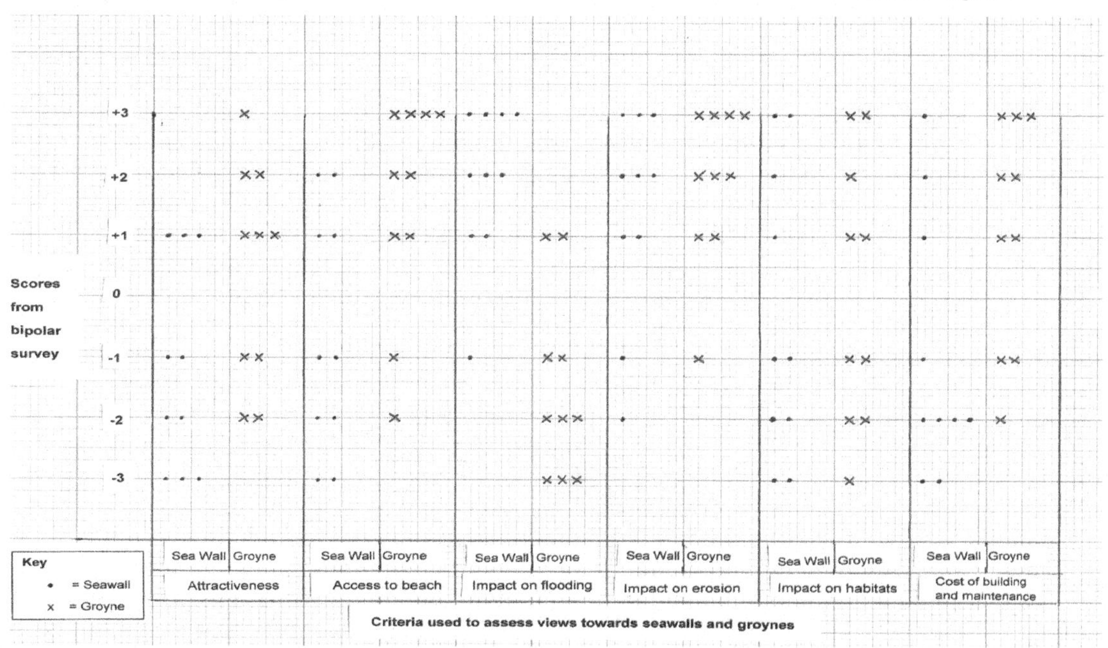

Activities

1. **Calculate the mean and median values for each of the six features for both sea walls and groynes. Use evidence from Figure 5.**
2. a) Suggest how you could present the results in Figure 4 as bar charts.
   b) Explain and justify whether you think the use of a dispersion graph or bar chart is a more effective way of presenting this data.
   c) Analyse the results of this survey. What conclusions can you reach about the attitudes towards sea walls and groynes as methods of coastal defence?

Questionnaires

Questionnaires are a great tool to collect up-to-date information of people's behaviours and views in relation to a particular issue.

Questionnaires often use a mixture of closed questioning (see page 72) and open questions. Open questions are those where the respondent is invited to answer in his or her own words and possible answers are not suggested. The table below shows a questionnaire which contains both open and closed questions. It was designed by some students to investigate how a new shopping centre might change shopping habits.

Figure 6 A draft questionnaire about the possible impacts of a new shopping centre

1. How old are you? ......

2. What form of transport do you use when you shop?

	Tick (✓)
Bus	
Car	
Bike	
Tram	
Taxi	

3. How far do you travel to do your shopping?

	Tick (✓)
0-5 miles	
5-10 miles	
10-20 miles	
20-30 miles	
Over 30 miles	

4. How often do you think you will use the new shopping centre?

....

5. What types of shop do you use most often?

....

6. Do you think that the new shopping centre gives better value for money and more choice than on the high street? **Yes/No**

7. Don't you agree that online shopping is better?

....

8. Do you think the town centre needs more parking? **Yes/No**

9. What benefits do you think the new shopping centre will bring to the area?

....

10. What problems may occur as a result of the opening of the shopping centre?

....

Designing the questionnaire

Use the following tips when posing the questions for your questionnaire.

- Ensure that the questions are relevant to your hypothesis/enquiry question.

- Consider whether you need to include both closed questions and open questions.
- Keep your question wording as concise as possible.
- Avoid jargon, for example, non-geographers may not understand the term CBD.
- Avoid leading questions. These are questions that lead the person you interview to agree with a view that you may hold, for example,

  'Don't you agree that the new cinema complex is a blot on the landscape?'.

- Avoid questions that ask two things in one go, for example,

  'Do you think that the new by-pass will bring jobs and reduce noise pollution?'

- Avoid 'negative questions', for example,

  'Do you disagree with the building of out-of-town shopping centres?'.

Carrying out the questionnaire

Step One Carry out a pilot survey (see pages 20-21) maybe on friends or members of your family. This will help you to identify and amend any potential problems with question wording.

Step Two Decide on your sample – which group(s) of people, how many people, location(s) of survey.

Step Three Choose a time and location where you are likely to be able to interview the number and type of respondents you wish to speak to.

Step Four If using more than one location, keep a record of where questionnaire surveys are conducted.

Dos and don'ts

Do:

- ✓ Introduce yourself and explain why you are conducting the survey and how the information will be used.
- ✓ Look business-like or professional by using a clip board and dressing smartly.
- ✓ Thank the respondents once they have answered your questions.
- ✓ Keep safe! Work in pairs, let people know where you are going and when you expect to return.
- ✓ Consider the reliability of drawing conclusions about the views of the total population based on a relatively small sample.

Don't:

- ✗ Don't get disheartened if people are unwilling to answer questions.
- ✗ Don't offend the respondent, for example, it is generally not a good idea to ask a person their age or how much they earn!
- ✗ Don't rely on your memory - record the responses immediately and clearly.
- ✗ Don't assume that all the responses you receive are completely accurate reflections of people's views, for example, the inaccuracy of opinion polls in recent elections.

Activities

1 Study Figure 6. Suggest how the wording and/or the structure of these questions could be improved.

Time to practise questions about qualitative data

Figure 7 A Likert survey designed to investigate the views towards the proposed expansion of a quarry in an area of high landscape value. It was conducted in a village close to the quarry

Gender		Age Group		
Male	Female	<20 years	20-60 years	>60 years

Where do you live? Record name of village or town (or nearest village or town)

____

For the following statements, choose the response which most closely matches your views.

A) I think expansion of quarrying will bring more jobs to this area.

Strongly agree	Agree	Neutral	Disagree	Strongly disagree
5	4	3	2	1

B) I think that the expansion of the quarry will provide much needed raw materials for local and national industries.

Strongly agree	Agree	Neutral	Disagree	Strongly disagree
5	4	3	2	1

C) I think that expansion of quarrying will deter tourists from visiting this area.

Strongly agree	Agree	Neutral	Disagree	Strongly disagree
5	4	3	2	1

D) I think that expansion of quarrying will create additional traffic problems in this area.

Strongly agree	Agree	Neutral	Disagree	Strongly disagree
5	4	3	2	1

E) I think that the quarry should be allowed to expand.

Strongly agree	Agree	Neutral	Disagree	Strongly disagree
5	4	3	2	1

1 a) Study **Figure 7**. Suggest why it could be useful for a survey about the impacts of quarrying to collect information about each of the following:

(i) the age of the respondents; [2]

(ii) the gender of the respondents; [2]

(iii) the place of residence of the respondents. [2]

b) Construct **two** more statements that you could add to the Likert survey about the possible environmental impacts of the quarry. [4]

2 Study **Figure 8**. It shows a photograph of an inner city area in the UK.

a) Suggest **two** annotations that could be added to the photograph to show how the environment might affect the lives of local people. [4]

Figure 8 An inner city area in the UK, Newcastle-under-Lyme

Resident responses

	6	5	4	3	2	1	Total	Mean Score
A: Housing	4 (x6)	2 (x5)	3 (x4)	1 (x3)			24 + 10 + 12 + 3 + 0 + 0 = 49	4.9
B: Litter	3 (x6)	2 (x5)	2 (x4)	2 (x3)	1 (x2)		18 + 10 + 8 + 6 + 2 + 0 = 44	4.4
C: Traffic		1 (x5)	2 (x4)	5 (x3)	2 (x2)		0 + 5 + 8 + 15 + 4 + 0 = 32	3.2
D: Pavements		3 (x5)	3 (x4)	2 (x3)	2 (x2)		0 + 15 + 12 + 6 + 4 + 0 = 37	3.7
E: Noise	3 (x6)	3 (x5)	2 (x4)	2 (x3)			18 + 15 + 8 + 6 + 0 + 0 = 47	4.7
F: Safety	4 (x6)	3 (x5)	1 (x4)	1 (x3)	1 (x2)			

Student responses

	6	5	4	3	2	1	Total	Mean Score
A: Housing		1 (x5)	2 (x4)	4 (x3)	3 (x2)		0 + 5 + 8 + 12 + 6 + 0 = 31	3.1
B: Litter		1 (x5)	2 (x4)	3 (x3)	3 (x2)	1 (x1)	0 + 5 + 8 + 9 + 6 + 1 = 29	2.9
C: Traffic			1 (x4)	3 (x3)	4 (x2)	2 (x1)	0 + 0 + 4 + 9 + 8 + 2 = 23	2.3
D: Pavements			4 (x4)	3 (x3)	3 (x2)		0 + 0 + 16 + 9 + 6 + 0 =31	3.1
E: Noise		2 (x5)	2 (x4)	2 (x3)	2 (x2)	2 (x1)	0 + 10 + 8 + 6 + 4 + 2 = 30	3.0
F: Safety			2 (x4)	2 (x3)	3 (x2)	3 (x1)		

b) Study **Figure 9.** It shows the results of surveys completed by 10 residents and 10 visiting students to the inner city area shown in **Figure 8**. Red numbers show the number of people who chose each score.

(i) Calculate the mean scores for both the residents and visiting students for statement F which is about how safe and secure the area feels. Show your working. [4]

(ii) What data presentation technique would you use to represent the results of this survey? Explain why you would choose this data presentation technique. [4]

(iii) Suggest **two** reasons why the views towards this residential area are different for the residents and visiting students. [4]

Figure 9 Responses collected by some students using the bipolar survey when visiting the inner city residential area shown in Figure 8

Chapter 10

Change over time

Learning objectives

In this chapter we will explore:

- how we can measure and present information about change over short, medium, and longer periods of time;
- how we can use secondary sources to assess changes in land use over time.

Investigating change over short periods of time

Some fieldwork enquiries focus on how a geographical pattern or process changes over time. You could investigate change over:

- very short timescales, for example, how the number of cars or pedestrians in a town centre varies during the day;
- medium timescales, for example, how the weather or discharge in a river varies over a number of days or weeks;
- much longer timescales, for example, how the number and type of shops varies from one year to the next.

Changing traffic flows

Evidence from traffic counts can be used to investigate questions about how flows of traffic change over time such as:

- Do diurnal (daily) variations in traffic flows indicate that a settlement is popular with commuters?
- Does increased traffic at weekends cause problems in popular tourist destinations?

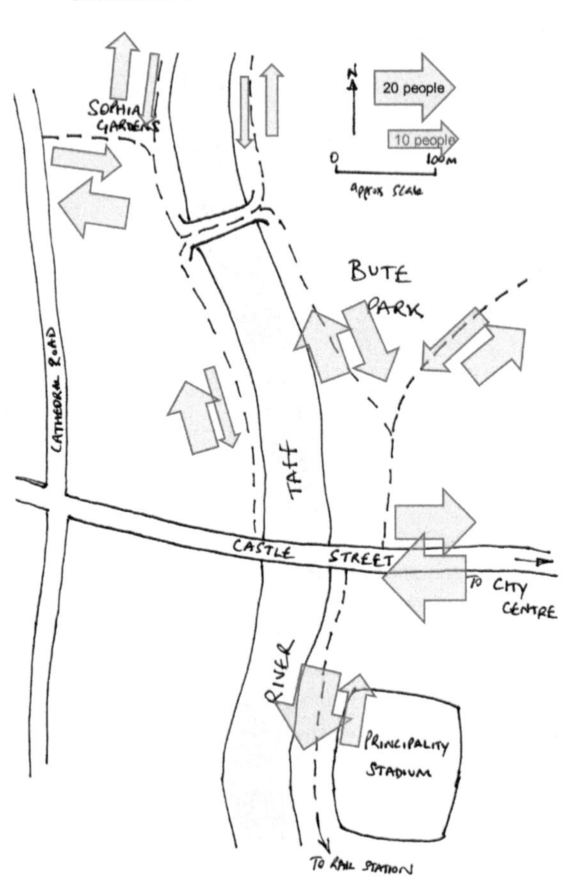

Figure 1 An example of a flow line map based on pedestrian counts

Section 2

Traffic and pedestrian surveys are explained in pages 56-57. When using traffic counts to compare change over time, it is important to:

- conduct the counts at the same locations as earlier counts;
- conduct more than one count to increase the reliability of results;
- take note of any factors, for example, differences in weather or changes to road layout which might affect reliable comparison between time periods.

Activities

1 Compare how pedestrian flows changed over time in Figure 1 and Figure 21 (page 40).

Investigating changes in the weather

Weather conditions in the UK change rapidly as warm and cold air fronts pass over. This can cause variations in temperature, cloud cover, and precipitation during a day. Primary data can be compared with secondary data from the Meteorological Office, or other national and local weather forecasts, to describe and explain changes. An example of a weather recording sheet is shown in Figure 2.

Figure 2 A weather data recording sheet.

Day	Mon	Tue	Wed	Thu	Fri	Sat	Sun
Maximum temp (°C)							
Minimum temp (°C)							
Rainfall (mm)							
Wind speed (km)							
Wind direction							
Cloud cover (eighths)							
Cloud type							
Air pressure (mb)							

Some equipment needed to collect primary weather data includes:

- an accurate thermometer to record temperatures;
- a rain gauge (made from a plastic bottle);
- a cloud recognition chart. Download an app to identify cloud types;
- a smart phone app to measure air pressure;
- an anemometer and compass to record wind speed and direction.

Strengths and limitations of collecting weather data	
Some strengths	**Some limitations**
Data can be recorded at systematic intervals during each day and/or over several days. These measurements can be used to analyse how weather changes as air pressure changes. Primary data provides detailed current information about how local weather changes. There are many accurate sources of secondary data to support investigations about weather.	Some pieces of measuring equipment, for example, anemometers are expensive but perhaps could be borrowed from school.

Investigating patterns of river discharge over time

River discharge is the volume of water which flows through a river channel in a given amount of time. It is usually measured in cubic metres per second (cumecs). River discharge is calculated by multiplying the river velocity by the cross-sectional area of the channel.

River discharge = cross-sectional area (m^2) x velocity (m/s) = (m^3)/s (cumecs).

In GCSE Geography, cross-sectional area can be calculated as:

Width of the river channel (m) x *average mean* depth (m) = m^2.

River depth and discharge can both vary a lot over relatively short periods of time. After heavy rain or snow melt, for example, river levels will rise. It takes a little time for water to move through the drainage basin and enter the river, so there is a lag in time of several hours between the peak rainfall and the peak discharge. Of course, it is not safe to collect discharge data in a river that is in flood. However, we can investigate how a river behaves during a flood event by:

- Looking for physical evidence that the river has flooded recently. This will take the form of rubbish – often plastics – caught in the branches of trees close to the river. The height of this rubbish above the river bank can be measured with a tape to give an idea of the level of the flood water.
- Mapping features in the urban environment that indicate that the river has flooded previously. Older buildings may have steps up to the front door. Modern buildings may have fixings outside for demountable defences to be slotted in.
- Collecting secondary evidence from websites about rainfall and discharge. Some websites publish historic discharge data. Others publish hydrographs, like Figure 1 on page 43, that show river levels for the last week or month. If this data is not available for your own river, then you should select data for a similar, nearby river.

Figure 3 Rubbish in the branches of a tree next to the River Lune (Lancashire) in March 2017. It shows that the river levels were several metres above normal during the flood in December 2015

https://www.riverlevels.uk/ This website publishes discharge data for rivers in the UK. You can use it to see how the discharge of a river changes with time. It is too dangerous to collect river velocity data when the river is in flood, so this website is particularly useful for seeing how river discharge changes after heavy rainfall.

How to collect data to calculate the cross-sectional area of a river channel

To calculate the cross-sectional area of the channel the width and average depth of the river channel must be measured.

Equipment needed includes:

- a tape measure;
- a metre rule;
- a data collection sheet to record the results.

Four people are needed: two to hold the tape measure; one to measure the river depth and one to record measurements.

Step 1 Two people hold the tape measure tightly across the river channel. The distance across the channel is measured and divided by 11 to give 10 equidistant depth measuring points, for example, if the distance across the channel was 4.4 metres, a depth measurement should be taken every 4.4m/11 = 40cm.

Step 2 Measure the depth with a metre rule. Try not to disturb material on the channel bed.

Step 3 Record the readings immediately and accurately. Use the same units of measure (usually metres for width and depth).

The size of the river channel and speed of flow must be considered to ensure it is safe to do fieldwork. Adult supervision is required.

Figure 4 This new house in Shrewsbury has been built above flood level. Notice the steps up to the front door that tell us how high the River Severn has flooded in the past.

How to compare changes in river discharge over time

To compare river discharge data with a previous week, month or year, data could be collected at different times and then compared. This data may be available within school from previous fieldwork investigations. Alternatively, primary discharge data can be compared with secondary data published on the internet. For some rivers, river level data may be obtained from the Environment Agency (apps.environment-agency.gov.uk/river-and-sea-levels) or from the website riverlevels.uk.

Primary data on river velocity, width and depth should be collected as described on pages 58 and 66. This data can then be used to assess how the river discharge has changed over time.

Strengths and limitations of measuring and comparing river discharge over time	
Some strengths	**Some limitations**
Accurate secondary data may be available from professional organisations to support your own findings. Additional evidence, for example, rainfall amounts can be analysed to make sense of changes in river discharge. It may be possible to explain changes in discharge data by analysing evidence of land use changes or changes in river management strategies.	Different types of equipment may have been used to collect the primary and secondary data, thus reducing reliability of comparison. To collect the information on width and depth and velocity, several students will need to be involved in the fieldwork.

Using photographs to analyse change over time

One simple way to see how much a place has changed over a longer period of time is through photography. Our towns and cities change a lot when redevelopment takes place. Old industrial districts change as factories and warehouses are demolished and new buildings - often shops, flats, and offices - go up in their place. High streets change when shops change hands, or when traffic is banned to make more space for pedestrians. These changes usually take place over several years so you will need to compare today's urban environment with secondary data to be able to analyse the changes. A great source of secondary data is old photographs, especially old postcards. These are available, to view or buy, on online auction sites. Once you have found some old photographs of your town you need to find the spot where they were taken. Then re-photograph the same place as it looks today.

Figure 5 Queen Street in Cardiff in 1946

Figures 5 and 6 show the same street in Cardiff. Today, this street is one of Cardiff's main pedestrianised shopping streets. Notice how much the street has changed for pedestrians. You can also see that several buildings have been replaced. We can analyse this type of evidence by annotating the photographs. Annotations are more complex statements than simple labels. A good annotation would compare the two photographs. It might go on to suggest why the changes have happened.

Activities

1 **Study the features that are numbered on Figure 6. Write annotations for these features.**

Figure 6 Queen Street in 2017

Section 2

How can primary and secondary data be used to assess changes in retail patterns?

Many of the shops and services in our cities, towns, and villages show gradual change year-by-year. Derelict sites are developed; shops change hands; and other shops and buildings become empty. The environment of each shopping street can also change. Some streets become pedestrianized; signage and benches for shoppers are improved; or parking restrictions are introduced. By revisiting a location that was surveyed in a previous year these changes can be analysed. Primary and secondary data that could be used to compare changes in retailing over time includes:

- pedestrian counts and traffic counts (see pages 56-57);
- questionnaires (see pages 76-77);
- environmental surveys (see pages 20-21);
- land use maps.

To be able to reliably compare results of pedestrian surveys taken in previous months or years, it is important to:

- conduct any counts at the same locations and times of day;
- use the same questions in your questionnaire survey as those used in previous fieldwork;
- check what sampling strategies were used and, if possible, use the same strategies;
- make a note of any differences in factors such as weather conditions, layout of buildings and roads.

Figure 7 Google Street View of Burslem in Stoke. The sliding bar shows that 6 sets of images are available, starting in 2009 and ending in 2016.

How can Google Street View be used to assess changes in retail patterns?

Google Street View can be used to carry out a virtual fieldtrip. The date that the photographs were taken is in the top left of the screen. Some shopping streets have been photographed 5 or 6 times by Google since about 2009. By comparing this secondary data to the primary data you collect, you can see which shops have closed or changed hands since the 'Street View' images were last taken. The change in the percentage of two types of shops is particularly interesting:

- empty (or vacant) shops – because an increase in vacant shops may indicate that fewer shoppers are using the town;
- locally owned (or independent shops) - because a decline in independent shops may indicate that the town is losing some of its identity and becoming more of a clone town.

How can maps be used as secondary sources to assess changes in land use over time?

There are numerous secondary sources of maps that can be used to investigate changes in land use over time. These sources include:

- maps drawn using data from previous fieldwork;
- local councils, libraries, or even tourist information offices;
- Digimap has OS maps, geological maps, and historic land use maps. Ask your teacher if your school uses Digimaps;
- GOAD maps are detailed street maps which include information about individual buildings and their use;
- Google maps can also be used to identify how current land use has changed in recent years.

Strengths and limitations of using maps as secondary sources	
Some strengths	**Some limitations**
Some maps can be used interactively and can be annotated in relation to the aim of your research. Tracing overlays can be used to add relevant annotations or diagrams to the map. Historical maps may no longer have copyright restrictions on use.	Some maps, for example, GOAD maps can be quite expensive.

To identify changes in land use it is useful to conduct a land use survey using a base map on which individual buildings are shown.

To conduct a land use survey, take the following steps.

Step One Use a base map such as a 1:1250 or 1:2500 OS map, if available.

Step Two As you walk around the area, mark on the map the type of land use, using a key.

If base maps are not available, you can make your own land use survey by noting the land use on either side of the road and recording it in two columns.

The categories in the key should relate to the aim of your investigation, for example, if you are investigating the impact of tourism over time, it would be important to include the type of buildings and maybe also tourist signage, litter bins, and car parks.

Step Three If possible, take photographs of land uses or events which are relevant to your investigation.

Step Four Once you have completed the survey, put the information onto a new map, using colours to show the different types of land use.

Dos and don'ts	
Do:	**Don't:**
✓ Always make sure that any map you use has a scale line, north arrow, and key.	✗ Copy a map from a secondary source without providing an accurate reference showing where you found the source, for example, the URL for a map you have found on the internet.

Time to practise questions about change over time

1 Students collected weather data over a 30 hour period. During this period of time a weather depression passed across their survey point.

Figure 8 Weather data collected as a winter depression passed over an area of the UK in January 2012

Time	Temp °C	Wind speed	Wind direction	Rainfall mm	Air pressure mb
6am	7.2	20.8	W	0.3	998
8am	8.0	17.2	SW	0.8	995
10am	10.3	19.9	SW	0.3	993
12pm	10.9	24.6	SW	0.5	992
2pm	10.9	32.2	SW	3.3	990
4pm	9.8	29.4	W	5.0	988
6pm	8.9	23.7	W	0.5	983
8pm	3.2	38.8	W	11.0	987
10pm	4.6	20.8	W	0.3	988
12am	3.7	26.5	NW	0.8	993
2am	4.3	21.8	NW	0	997
4am	4.6	24.6	NW	1.0	1002
6am	4.0	18.9	N	0	1006
8am	3.2	12.3	NW	0	1008
10am	5.2	14.2	NW	0	1010
12pm	6.0	18.9	NW	0	1012

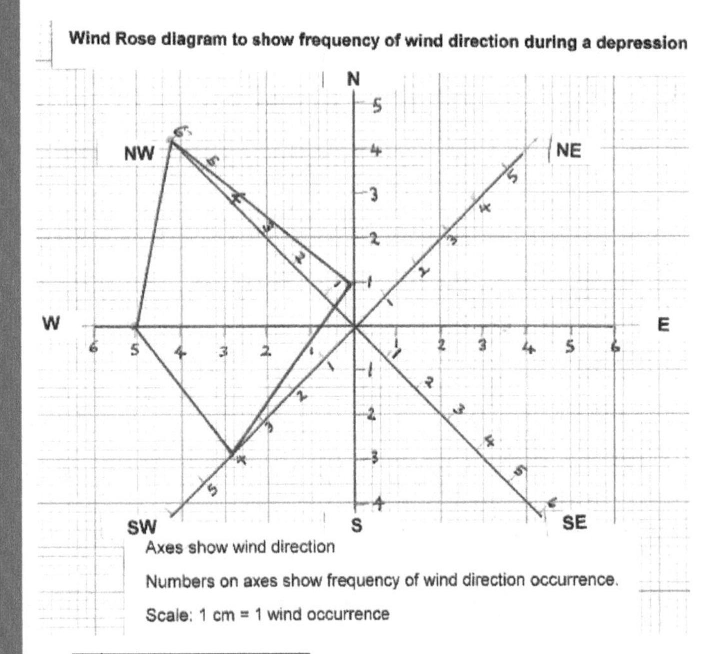

Figure 9 Wind rose diagram to show frequency of wind direction during the passage of the depression

a) Study **Figure 8**.
   (i) Suggest how the data on temperature and rainfall could be presented as a graph. [4]
   (ii) How does temperature change as the depression passes over? [2]

b) Study **Figure 9** below.
   (i) What does **Figure 9** show about dominant wind directions during the depression? [2]
   (ii) What does **Figure 8** suggest about how wind directions change as the depression passes over? [2]

c) Explain whether the information in **Figure 8** and **Figure 9** supports the idea that depressions bring periods of changeable weather. [4]

2 Students collected data about a river channel. Their data is shown in Figure 10. They intended to compare this data with secondary data showing how the discharge of a nearby river changed over a month before the day of their fieldtrip.

Site name (from upstream to downstream)	Width (m)	Average depth (m)	Cross sectional area) m³	Velocity (m/second) Measured using flow meter	Discharge (cumecs)
A	1.68	0.06	0.10	0.34	0.03
B	2.10	0.06	0.13	0.40	0.05
C	1.35	0.08	0.11	0.34	0.04
D	1.10	0.18	0.20	0.50	0.10
E	2.60	0.07	0.18	0.39	0.07
F	6.00	0.34	2.04	0.60	1.22
G	5.60	0.38	2.13	0.81	1.72
H	5.70	0.37		0.73	
I	6.00	0.44	2.64	0.76	2.00
J	8.28	0.29	2.40	0.85	2.04

Figure 10 Some results of river measurements on a small river in Cumbria

a) Study Figure 10.

(i) Use the width and depth measurements for site H to calculate the cross-sectional area of the river at this site. Show your workings. [2]

(ii) Calculate the discharge at site H by multiplying your answer to task (i) by the figure for velocity at this site. [2]

(iii) Suggest how dredging could change the cross-sectional area of the channel. [4]

(iv) Explain why removal of trees in drainage basins could change the river discharge. [4]

3 Study Figure 11. Use evidence in the photograph to suggest how you could investigate:

a) Differences between weekday and weekend tourism. [4]

b) Changes to retailing over time. [4]

Figure 11 Ironbridge, a honeypot site in Shropshire.

SECTION 03

Chapter 11 The concept of Place

Each place has features that give that place a unique character. People recognise this character – it gives the place, and the people who live there, a sense of identity.

Learning objectives

- Understanding the concept of place and how it might be investigated through fieldwork.

What is this concept about?

A knowledge and understanding of place is one of the most important features of geography. In the first instance we need to be able to locate the place. We also need to find ways of describing the character of the place. This will include describing its physical nature and its human features. Finally, to give the place a true identity we need to be able to understand the combination of things that makes it different from other places; factors such as the social make-up of its residents, its cultural, historic, and economic background.

How can fieldwork help us investigate places?

An important starting point is to decide the particular focus of the investigation. For example, you should decide exactly what aspect of the place you want to investigate. Some examples are:

- which features make a place unique;
- how and why a place changes over time;
- how the identity of a place is influenced by its links to other parts of the world;
- why a place attracts people to live in it, or visit it;
- how different people feel about the place (their perceptions);
- how the place tries to promote its image in the media.

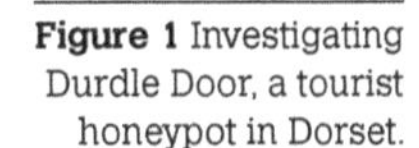

Figure 1 Investigating Durdle Door, a tourist honeypot in Dorset.

We could investigate the identity of this place by:

- collecting qualitative data about the physical, cultural, historical and social features that give this place a unique identity. Data collection methods could include a questionnaire of both local people and visitors, or analysis of internet sites and blogs that describe this place;
- taking photographs of local landscape features. These photographs could then be used in a Likert Survey to investigate whether locals and visitors have similar opinions about the importance of the landscape.

Figure 2 The identity of most towns in the UK is shaped by their links to the rest of the world. This may be due to migration or the process of globalisation. This Italian American restaurant is in Ipswich, UK.

Are towns losing their identity?

Across the UK, many people are concerned that towns are losing their identity. The fear is that locally owned independent shops, unique to each town, are gradually being replaced by shops owned by national chains.

The Royal Geographical Society has produced a survey sheet to help you investigate if your town is a Clone Town. Is it similar to towns around the country, or is it a genuine Home Town that is distinctive and recognisable as a unique place?

The Clone Town Britain Survey can be used in any town in the UK. You simply need to record the types of shop in the high street using the categories provided by the Clone Town Survey. Then use a formula to give a score out of 60. Scores below 25 suggest that the town is already a clone town; scores above 40 suggest that the town has a unique identity in terms of shopping. You can also compare your town with another in a different region.

Fieldwork investigating place can be supported by using Geobrowsers such as Google Earth. Geobrowsing technology gives a 3D computer generated representation of the Earth which enables you to view places and different aspects of the environment by rotating the Earth and changing the angle, as if you were flying over an area.

Activities

1. **Identify 5 ways that Geobrowsers would be helpful in the planning stage and also after the fieldwork has been completed. Compare your list with a partner.**
2. **Suggest a hypothesis that would form the starting point for fieldwork investigating three different aspects of place:**
   - a place attracting large number of people – a honeypot;
   - a place changing over time;
   - a place in danger of losing its identity.

We could investigate how a place is shaped by its global links by:

- surveying and mapping all of the retail and service provision in a town that has clear links with the wider world. The land-use map key could reflect any world links based on continents or countries.
- using secondary data sources, such as the census, to identify parts of the city in which people of different faiths, cultures or origins live. Asking people who live in these parts of the city to explain how their global links have shaped their view of their home town.

http://www.rgs.org/HomePage.htm **Go to the RGS home page and search for their clone town survey technique.**

Comparing perceptions of place

Tourists generally have a positive image of the place they intend to visit. Why else would they visit it? Internet sources and word of mouth may have created this positive view. The same tourist may also be aware of some negative features about the place. Such negative views might relate to traffic jams, high car park charges, or overcrowding. Fieldwork enables students to investigate the concept of place and how the reality might be different from the expectation. Pages 72-77 provide lots of ideas for collecting qualitative data on how the perception of a place might differ before and after a visit. Well-constructed bipolar surveys, questionnaires and Likert surveys would be ideal for a range of investigations at honeypot sites: Such investigations might include:

- how visits to honeypot locations change the way tourists perceive the place;
- whether different age groups have different perceptions of a tourist location - before or after a visit (or both);
- whether honeypots, famed for their landscape value, are more appreciated than built attractions (for example, large theme parks).

Different ways that people perceive a place

Fieldwork can be used to investigate differences in the way that places are perceived. Rural areas provide an ideal opportunity for investigating this. Tourists who like to visit quaint old rural villages may already have a fixed idea about the village before they arrive. This is worth testing if you interview them as they arrive in a car park. The visitors may have been encouraged to visit after seeing information in guide books or by viewing tourist board websites. If their visit is planned at a time when the sun is shining, and if there is a bonus of a village festival or fete, their initial perception about the place is likely to be confirmed when you interview them on return to the car park. On some occasions their perception may have changed. You would need to ask them why.

Figure 3 Festivals, like this one in Shropshire, attract visitors to rural areas

To investigate perceptions of place it is important to plan ahead. Perhaps you need to select a day when an event is taking place. You will certainly need to ask the right questions to the right people. Questionnaires or survey sheets need to be designed carefully to ensure that you are certain if those being questioned are locals or tourists. You will need to record this important information at the top of the survey sheet to help in the analysis phase. If you asked the tourists in Figure 3, it would be no surprise if they gave the view that the village was vibrant and a friendly place to live.

If you ask locals, you might record very different views. For those not involved in the festival, they might report that their business is disrupted by road closures and parking restrictions on the day. They might report other issues such as house prices rising due to newcomers wanting to live in the village. Other locals might welcome the boom in business if their business relies on tourism. These conflicting opinions confirm why it is important to record details about the people you are interviewing.

In Sections 1 and 2 of this book you were asked to consider the strengths of using open and closed questions in questionnaires. In the case of perception studies, asking open questions alongside some closed questions will allow those being surveyed to speak from the heart.

An example is taken from interviews carried out by geography students in the area surrounding Loch Lomond, a honeypot located 40 km north-west of Glasgow. Each student was asked to interview two people; a tourist visiting the area and an elderly resident who had lived in the area for seventy years. They recorded the responses on their mobile phone rather than try to write down everything that was said.

Figure 4 Different views about Loch Lomond

"The area is not what it used to be. The beautiful shore of the Loch is spoilt by rough campers who light fires and leave rubbish and human waste all over the place. You can no longer sit and enjoy the bird song as too many motor boats are speeding up and down. It's been made worse since they improved the A82. In the holiday period you can't find anywhere to park and the road around the Loch is always congested."

"This is a great place for letting off steam. I can be enjoying my jet ski within one hour of leaving Glasgow. The camping is a bit rough but there's nothing better than partying at a night-time BBQ with a few beers and friends watching the star filled sky. The locals are really friendly, even when you knock on their door asking for some water for cooking! I wouldn't mind living here sometime in the future."

Activities

1. **Give two reasons why it is important to collate responses from a range of people when interviews are used to learn how different groups of people identify with a place. Compare your ideas with a partner.**
2. **Suggest one open ended question that would be effective in asking an interviewee to say how they identify with a place. Compare your question with a partner.**
3. **Discuss the advantages and disadvantages of using mobile phones to record responses about perception of place.**
4. **Discuss the difficulties of analysing responses gained from using open questions.**

Strengths and limitations of perception studies	
Some strengths	**Some limitations**
Older residents are a rich source of information for students who want to investigate change or how they perceive a place. School links with sheltered accommodation or care homes are an ideal starting point.	Tourists may not want to be interviewed if they think it will take a long time. Choose your questions very carefully so you can collect useful information from short interviews.

Time to practise questions about fieldwork on place

1. Study Figures 5, 6 and 7. They show important features of the postcode district CF11 in Cardiff. The centre of this residential area is about 1.5km from Cardiff city centre.

Figure 5 The green space of Bute Park, next to CF11.

Figure 6 Welsh language is spoken by many in CF11.

Figure 7 CF11 has a multicultural population.

a) Students carried out a pilot survey to create a short list of features that people thought gave CF11 its local identity. They chose three features that came up in the pilot and photographed them. They then used a Likert survey. They asked local people to look at each photograph and asked people to respond to the following question:

Do you think that this feature is an important part of the local identity of this part of Cardiff?

(i) Suggest **two** advantages of using a pilot survey in this fieldwork enquiry. [4]

(ii) Suggest **one** limitation of the way the students organised this data collection. [3]

b) The students' aim was to discover whether younger residents had a significantly different opinion about local identity than older residents. They used the census to find information about the population of CF11. The results are shown in Figure 8.

Figure 8 Data from the Census for the post code district CF11 and post code unit CF11 7BH

Age range (years)	CF11		CF11 7BH	
	Population	Per cent	Population	Per cent
0-29	18,374	45.7	406	
30 +	21,784	54.3	110	
Total	40,158		516	

(i) Students decided to interview 100 people in post code CF11. They decided to interview 46 people aged 29 or younger. The rest would be aged 30 or over. What is the name given to this sampling technique? [1]

(ii) Suggest **one** advantage of this sampling technique for investigating opinions about local identity. [3]

(iii)Suggest **one** limitation of this sampling technique if you were working in post code district CF11. [3]

(iv)Calculate how many people would be interviewed if this sampling technique were used in CF11 7BH. [3]

(c)The students' results for their survey of CF11 are shown in Figure 9.

Age range (years)	Historic Bute Park is an important part of CF11's identity		Welsh language is an important part of CF11's identity		The multicultural population is an important part of CF11's identity	
	Agree	Disagree	Agree	Disagree	Agree	Disagree
0-29	21	79	75	25	80	20
30 +	55	45	52	48	78	22

Figure 9 Results from the students' survey

(i) Suggest **one** way that you could represent the data in Figure 9 graphically. Justify your choice of presentation method. [3]

(ii) What conclusions can you reach from this data? [4]

2. Study Figure 10. It shows shops in the High Street of Shrewsbury. Students wanted to investigate the factors that give Shrewsbury town centre its local identity.

Figure 10 High Street in Shrewsbury.

a) Identify **two** features that might create a local identity for this place. Use evidence from Figure 10 only. [2]

b) The students decided to investigate whether Shrewsbury is a clone town. They collected data about types of shops in the town centre. Some of their results are shown in Figure 11.

Postcode	Mean rateable value (£ per m²)	Number of shops in this post code that are:		
		Independently owned	Chain stores	Vacant
SY1 1SZ	225	7	0	2
SY1 1SP	353	3	14	1
SY1 2BU	383	0	9	1

Figure 11 Data on shop types and rateable values in three different streets in Shrewsbury town centre.

(i) What conclusions can you reach from this data? [4]

(ii) Explain **one** limitation of this dataset. [2]

(iii) Mean rateable value was found from a secondary data source. Give **one** reason why this data was helpful in this enquiry. [2]

Chapter 12
The concept of Sphere of influence

Geographical features have positive and negative effects on the environment surrounding them. This area around a feature is called its sphere of influence. The nature of the feature influences the size and impact of the sphere of influence.

Learning objectives

- Understanding the concept of sphere of influence and how it might be investigated through fieldwork in a variety of environments.

What is this concept about?

The features in our environment can have good or bad impacts on us. Compare Figures 1 and 2. Parks and busy roads are common features of the UK urban environment. Think about the impacts of the park. People find it attractive. It provides a safe place for exercise and relaxation. The good features of the park have a positive impact on the lives of local people. Now think about the impacts of the busy road. It does have some positive impacts for local people. It means locals are well connected to other places, especially if it has a bus route. However, the noise of the traffic, and the fumes they emit, will have negative impacts for some local people. Notice that we could categorise each of these impacts as impacts on local people, economy, or environment.

Figure 1 A public park in the centre of Cardiff, Wales

We could investigate the sphere of influence of the park by:

- asking people what they think of the park using a Likert Survey. Use strong statements like 'It would be a great benefit to live close to this park'. Survey people to see if they agree or disagree;
- use local estate agents to find local house prices. If the park really is attractive to people, then we might expect that houses overlooking the park to be more expensive than similar houses a few streets away.

Figure 2 Traffic on a busy road in Newcastle-under-Lyme

We could investigate the sphere of influence of the road by:

- taking noise readings – perhaps using an app on your smart phone. How quickly do noise levels decrease as you move away from the road? Do noise levels vary throughout the day?
- measuring the amount of dust in the atmosphere by sticking double-sided tape to buildings to collect dirt and dust on the open sticky side. Is there more dust close to the road?

Are the impacts mainly good or mainly bad?

Some features in our environment have mainly good impacts. An example would be the flood defences along the banks of a river, or a sea wall in a coastal town. You could say they have been designed to have a beneficial sphere of influence by saving lives and property.

Other features have mainly good economic impacts but bad impacts on the surrounding environment. Airports, for example, create a lot of jobs in the local economy. However, they create noise nuisance as aircraft take off and land. So, people might complain about living within their sphere of influence. We get another concept from this known as NIMBYism. This occurs when people protest because they don't want a particular development (like a new runway) built close to where they live.

Figure 3 Flood gates at Sea Palling in Norfolk have a positive impact on homes and businesses in the flat, low-lying coastal areas behind this coastal defence

We could investigate the sphere of influence of these flood defences by:

- using an online flood risk map to identify places that are vulnerable to flooding and other places that are protected by the defences;
- using transects at right angles to the coast and identifying land uses that are vulnerable and land uses that are protected.

Activities

1 **For each of the following features, suggest how you might investigate its sphere of influence. Remember, your investigation would be looking for evidence of good and bad impacts of the feature on its surrounding area.**

a) A large out of town supermarket.

b) A beauty spot in a National Park.

http://www.checkmyfloodrisk.co.uk/
https://www.naturalresources.wales/our-evidence-and-reports/maps/flood-risk-map/

These sites provide interactive maps about flood risk on rivers and coasts.

Learning objectives

- Understanding that the impacts of a feature decrease over distance.
- Considering how to investigate the positive and negative impacts of an event.

The effect of distance

We have seen that features of the environment have impacts that can be good, bad, or a mixture of both. These impacts spread out – rather like ripples on a pond when you throw in a stone. These are the 'spheres' of the spheres of influence. Some impacts are only felt close by. Other impacts extend much further than others. All of the impacts tend to decrease over space – an effect that is called **distance decay**.

Distance decay could be investigated through fieldwork. For example, how far:

- does litter spread from a take-away restaurant?
- do parking issues spread from an event such as a football match?

Study Figure 4. It shows a large building site in central Cardiff. You could investigate its impacts on the surrounding area by designing an Environmental Quality Index (see pages 20-21). Students used this EQI to measure noise from the building site along two transects. Each transect started at the building site. Notice how noise levels decrease with distance on the finished graph.

Figure 4 Investigating distance decay of noise nuisance from a building site in Cardiff

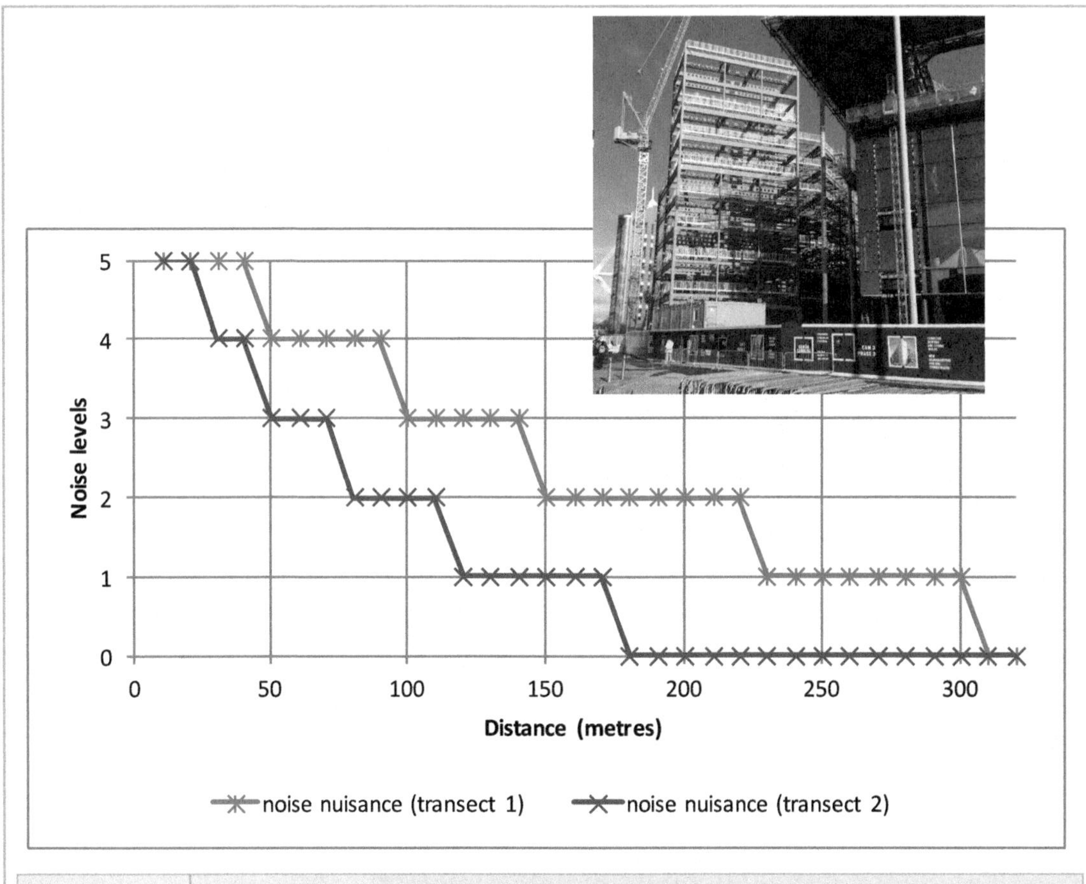

	Criteria				
	5	4	3	2	1
Noise levels from the building site	Peak noise from site much louder than traffic.	Noise from site similar to level of traffic noise.	Noise from site a little quieter than traffic noise.	Only heard as a background noise.	Only heard when there is no other noise.

Activities

1 **Study Figure 4.**
   a) Compare the way in which noise nuisance declines along each transect.
   b) Suggest one possible reason for the differences.

2 **Think about each of the following features of the environment.**
   - Airport
   - Take-away restaurant
   - Nuclear Power Station
   - Stone quarry
   - A beauty spot in a National Park
   - Large out-of-town supermarket
   - Premiership football ground

   Discuss how extensive you think the sphere of influence might be. You could use a scale from small to large:
   - Very local (a few metres)
   - Local (a few 10s of metres)
   - Moderate (for example, a neighbourhood in a town)
   - More widespread (for example, a town)
   - Large scale (for example, a county or region)

3 **Research an event that takes place in your locality. Discuss how you could investigate its sphere of influence.**

Investigating spheres of influence of an event

Sporting events, music and arts festivals and even village fetes can all have impacts on the surrounding environment and local economy. Positive impacts can be created when visitors spend money in local shops, cafes, pubs, or hotels. However, events and festivals can also create negative impacts such as noise, traffic congestion, parking issues, or anti-social behaviour.

Figure 5 The Principality Stadium in Cardiff is used for international rugby and football matches. The stadium is also used as a music venue

We could investigate the sphere of influence of an event by:
- comparing traffic congestion, parking issues and flows of pedestrians on the day of the event with a day when the event is not taking place;
- using a questionnaire to ask local people and visitors about the impacts of the event. Does one group of people have a more positive view of the event than the other group?

Time to practise questions about fieldwork on spheres of influence

Some students decided to investigate the sphere of influence of an event in a small rural town. The event was called the 'mid-summer fair' and took place over one weekend in June. Visitors were attracted to:

- **a classic car and traction engine parade in the main street;**
- **a fun fair for children;**
- **live music that was staged in a tent in the town square.**

Figure 6 The classic car and traction engine parade in the main street of Bishop's Castle, Shropshire

1 The students wanted to use a hypothesis.

a) Study the following hypotheses. Which **one** hypothesis could be used to investigate **spheres of influence**? [1]
- *'The mid-summer fair uses the rural identity of the town to attract visitors.'*
- *'Noise levels in the town were higher during the fair than during normal weekends.'*
- *'More families attended the fair than single people.'*

b) Suggest another hypothesis that could be set about either a positive or negative impact of the fair. Use Figure 6 to help you. [1]

c) Give **one** reason why your hypothesis is suitable for an investigation of this concept. [2]

d) Another student suggested this hypothesis:
- *'The summer fair was a huge success.'*

Give one reason why this is not a suitable hypothesis for an enquiry about the concept of sphere of influence. **[2]**

2 About 2,500 people live in the town and about 1,250 visitors came to the event. The main street was closed to traffic for the whole weekend. The students decided to investigate the impact of the road closure on local people and visitors. They used a Likert Survey. They asked 10 local people and 50 visitors. The details of their sampling strategy are shown in the table below.

	Total number	Number of people in the sample
Local people (who live in the town)	2,500	10
Visitors to the event	1,250	50

a) If the students had asked 30 local people, and used a stratified sampling technique, how many visitors should they have asked? Show your workings. [2]

b) Give **one** reason why a stratified sample would have given more reliable results than that used by the students. [2]

3 The results of the Likert Survey are shown in Figure 7. The students used Figure 8 to present these results.

'The closure of the main street to traffic was a good thing'	**Number of responses from ...**	
	Local people	**Visitors**
Strongly agree	0	10
Agree	1	30
Neutral	1	7
Disagree	5	2
Strongly disagree	3	1

Figure 7 Likert Survey results. People were asked to respond to this statement: 'The closure of the main street to traffic was a good thing'.

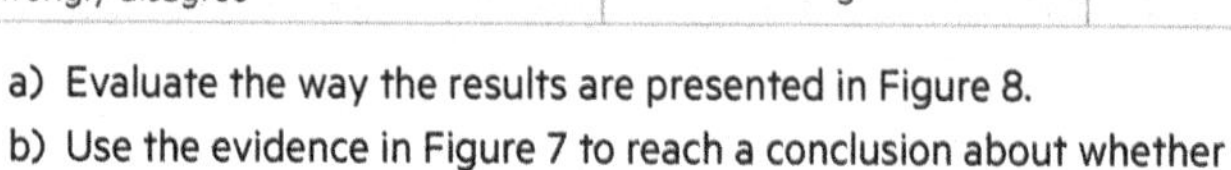

a) Evaluate the way the results are presented in Figure 8. [4]

b) Use the evidence in Figure 7 to reach a conclusion about whether *'The closure of the main street to traffic was a good thing'.* [2]

c) Suggest **two** reasons why some locals may have disagreed with the statement. [4]

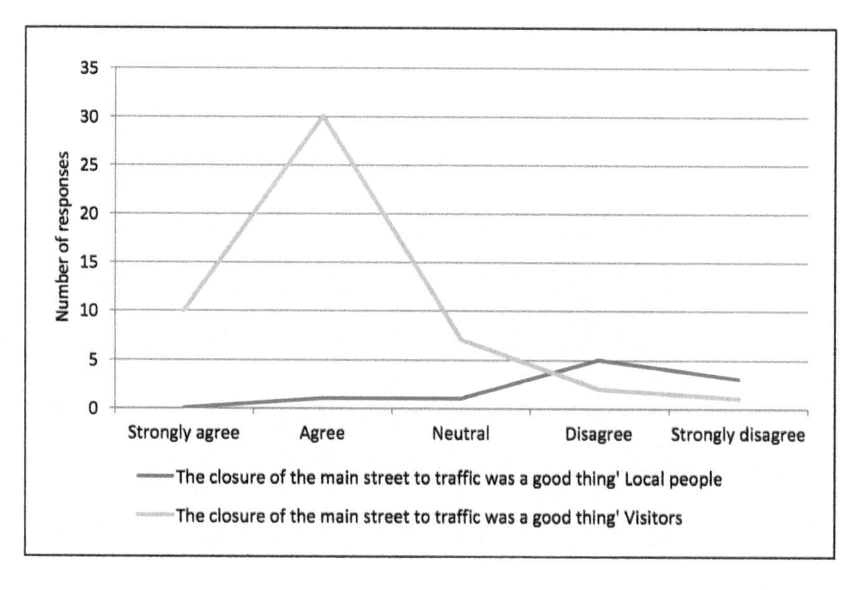

Figure 8 Line graph showing the Likert Survey results. 'The closure of the main street to traffic was a good thing'.

Chapter 13

The concept of Cycles and Flows

Environments contain flows, for example, water, people, and money. These move between stores. This concept explores the factors that influence the size and rate of these flows.

Learning objectives

- Understanding the concept of cycles and flows and how it might be investigated through fieldwork.

What is this concept about?

Cycles and flows are two related concepts that help us to explain patterns of movement. Pages 56-61 describe how flows can be measured. In this chapter, we will explore how the concepts of cycles and flows can help us make sense of geographical processes and patterns – using the concepts to explore the factors that influence the size and rate of these flows.

Investigating flows in the water cycle

In the water cycle, flows of water move between different places where water is stored. Water is stored in the atmosphere, or in surface stores, vegetation, soil, and groundwater. Water flows between these stores by evaporation, precipitation, drip flow, surface runoff, and throughflow. River fieldwork is a great place to investigate how these stores and flows are connected to one another.

We could investigate the concept of cycles and flows in a river by:

- measuring river discharge and velocity at significant distances down the long profile of the river. Then using the iGeology app (see pages 24-25) to investigate whether changes in flow can be explained by the rock type of the catchment area;
- measuring river velocity (see page 58) and cross-sectional area to calculate discharge. Then comparing this to rainfall or discharge data using secondary sources of data (see pages 26-27).

Figure 1 Investigating the relationship between stores and flows in a river. Cardingmill Valley, Shropshire

An enquiry located at a honeypot site

Ecosystems demonstrate a particular type of cycle that develops over long periods of time. Plants are replaced by other species of plant over time. Eventually a collection of plants develops that is typical for that environment. Trampling through the vegetation damages plants and can modify this natural cycle.

Figure 2 Investigating flows at a honeypot site. Stiperstones, Shropshire

We could investigate the impact of visitors on natural cycles at a honeypot site by:

- investigating the impact of visitor trampling by recording plant type and height. You would do this by using a quadrat (see pages 14-15);
- counting the number of walkers at different distances from the visitor centre / car park and comparing this to the width of the footpath.

A coastal enquiry

In a coastal environment, sediment is transported by the action of waves and currents. As the waves and currents slow down they lose energy. As they do so they begin to deposit the sediment. The largest, heaviest sediment is dropped first. Smaller and lighter sediment is transported further and deposited later. This process is known as sorting. So, by investigating the pattern made by sorting we can come to conclusions about the direction of flow of water in the coastal environment.

Figure 3 Investigating the movement of sediment up and down the beach - Borth, Wales

We could investigate the concept of cycles and flows up and down a beach profile by:

- using a clinometer to measure the beach profile (see pages 16-17). Then, using quadrats, sampling the size and shape of the beach sediment along the profile;
- collecting litter from different parts of the beach profile. You could analyse the litter by sorting (plastics, tins, and polystyrenes) and weighing the types.

Activities

1. **Students wanted to investigate the factors that influence discharge as a river flows downstream. Describe how you would collect data for each variable using a mixture of primary and secondary evidence.**
2. **A student set the following hypothesis:**
   'Pebbles have been sorted by size along the length of the beach'.
   a) Describe the sampling strategies you could use to test this hypothesis.
   b) Explain why it would be useful to measure the strength and direction of the prevailing wind **on more than one occasion** to support the evidence of this investigation.

Investigating flows of people in an urban environment

Our towns and cities also contain flows and stores. Traffic, commuters, goods and services all flow in and out of urban environments. We can also identify stores – places where people and traffic pause for periods of time before moving on again. These stores could be the offices where people work, the shops, parks and cafes we visit, or the car parks where we leave our cars.

Investigating transport management

Flows in an urban environment involve very large numbers of people and traffic. So, urban areas are sometimes congested, especially at key times of the day and week when commuters are trying to get to and from work. Traffic and pedestrian flows need very careful management so that they are both safe and efficient. Planners try to reduce travel times by developing integrated transport systems – where public transport, cycle lanes and safe pedestrian routes are all carefully connected. Planners also help to reduce risk by introducing traffic calming measures and/or trying to separate pedestrians from traffic. These plans make our cities safer and more sustainable. Your fieldwork could investigate how people manage transport systems to try to control flows of pedestrians, cyclists, or traffic.

We could investigate the management of transport in an urban area by:

- deciding on the best route for a new cycle path by designing an Environmental Quality Index (see pages 20-21) which assesses the attractiveness, safety and practicality of three different routes through a city;
- mapping traffic calming measures such as pelican crossings, speed bumps, or pinch points. Then asking local residents or pedestrians whether they are satisfied with efforts to manage traffic flows in the neighbourhood by using a questionnaire, or Likert Survey.

Figure 4 Investigating the management of transport systems - Riverside, Cardiff

Activities

1 **Use secondary data to investigate the pattern of rateable values across your town centre. Are rateable values highest on streets that have most pedestrians?**

a) Find postcodes for different parts of your town centre by using Google Maps and opening links to businesses on the map.

b) Use the postcodes to search the rateable value database. Rateable values vary a lot because they depend on the size of the shop so it is best to record this as a value per square metre.

c) Represent this secondary data on a sketch map of your town. You could use located bar charts. What pattern can you see?

Investigating the location of retailers in an urban environment

Shop owners (retailers) want their shops to be easily accessible to their customers. This means that retailers are interested in factors such as:

- whether there is sufficient parking nearby. If not, is there a park and ride scheme?
- the quality of the shopping environment. Will their customers be satisfied with the quality of pavements, street furniture, or signage that directs them around the town centre?

Most of all, most retailers in our town centres want their shops to be in locations that have high footfall – this simply means places where there are high volumes of pedestrians. These locations tend to be highly desirable. Consequently, shop owners tend to pay a higher rate of tax (known as the rateable value) for these shops. This relationship may explain why many small, locally owned businesses can no longer afford to locate in the busiest streets of our town centres. Fieldwork would allow you to investigate a hypothesis such as *'Rateable values are highest in streets that have the highest footfall'* or *'Footfall is highest in streets that are dominated by chain stores'*.

This web link includes a link to an online tool that allows you to find the rateable value of shops in your high street.

https://www.gov.uk/correct-your-business-rates

Figure 5 Investigating footfall and its relationship to shop type in a retail environment - Cardiff

We could investigate whether the location of retailers is affected by flows and cycles of pedestrians in this retail environment by:

- comparing primary evidence of pedestrian flows to the rateable values to see whether you can explain why large chain stores are located in one part of the town centre whereas smaller, independent shops are found in other parts of the town;
- using the bipolar technique to record perception of positive and negative features of the urban environment (for example, busy roads, safe places to cross, available parking, attractive street furniture). Do these likes and dislikes help explain why some retail locations have more pedestrians than others?

Time to practise questions about fieldwork on cycles and flows

1 **A group of students collected data on pebble size at four locations along the beach shown in Figure 6. They wanted to investigate whether sediment was sorted by size as it was transported along the beach. Longshore drift transports pebbles towards the position of the camera. Their data is shown in Figure 7.**

a) Study Figure 7. For each quadrat:

(i) Calculate the range. [4]

(ii) Calculate the interquartile range. [4]

Figure 6 Students used quadrats to collect data about pebble size along this beach in Vik, Iceland

Figure 7 Length of pebbles (mm) at four locations along the beach

Length of pebbles (mm)			
Quadrat A	Quadrat B	Quadrat C	Quadrat D
11	95	18	25
10	60	17	22
10	35	17	20
9	30	16	20
9	20	15	18
9	18	14	15
8	14	13	14
7	14	11	14
7	11	11	12
6	10	10	11
6	9	10	10
6	9	9	9
5	8	9	9
4	8	7	7
4	7	7	8

b) The students have muddled up the quadrats. They cannot remember in which order the data was collected.

(i) State the order in which you expect the quadrats were collected, starting with the one furthest from the camera. [4]

(ii) Explain why you have chosen this order. Use evidence from Figure 7 to support your answer. [4]

2 Another group of 8 students investigated the retail environment of Shrewsbury. They wanted to investigate whether footfall was highest outside shops that had the highest rateable values. They worked in pairs and collected data at ten locations across Shrewsbury. These locations are shown on the sketch map in Figure 8. Some of their data is shown in Figure 9.

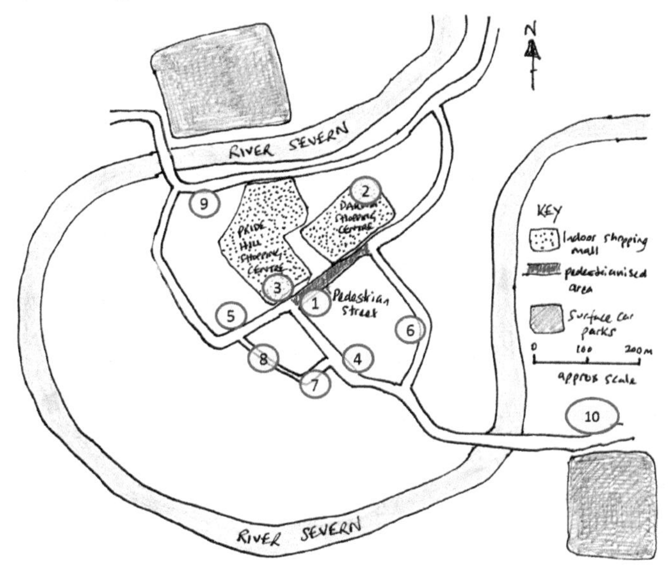

Figure 8 Sketch map of Shrewsbury

Figure 9 Rateable values and footfall

	Postcode	Mean rateable value (per m²)	Footfall (pedestrians in 2 minutes)
1	SY1 2BU	720	68
2	SY1 1PL	648	72
3	SY1 1DB	383	85
4	SY1 1SP	353	70
5	SY1 1HS	350	45
6	SY1 1EN	232	24
7	SY1 1SZ	225	15
8	SY1 1LW	225	12
9	SY1 1PP	218	24
10	SY2 6AE	130	15

a) What conclusion can you reach from the data in Figure 9? [2]

b) Suggest one suitable mapping technique to represent the data in Figure 9. Justify your choice. [4]

c) Suggest one suitable technique to represent the correlation in Figure 9. Justify your choice. [2]

d) Suggest two different reasons for your conclusion. Use evidence from Figures 8 and 9 to support your answer. [4]

e) Suggest one major limitation of the data collection technique used by these students and one way in which this could have been improved. [4]

Chapter 14

The concept of Mitigating Risk

All environments contain risks. These risks can be reduced (or mitigated) by managing the hazard or by changing the behaviour of people who may be affected by them.

Learning objectives

- Understanding the concept of mitigating risk and how it might be investigated through fieldwork.

What is this concept about?

Everywhere we go we face risks and hazards. Compare Figures 1 and 2. Town centres present us with a wide range of hazards. Some risks are obvious, such as those linked with the busy traffic we experience in town centre locations. Some risks are less obvious, such as the potential for street crime at night. The coastal areas we visit are also full of potential risks. We might seek out a beautiful viewpoint at the top of a cliff. To enjoy the best secluded beach we might risk being cut-off by the incoming tide. As individuals we weigh up the risks and hazards we face. We change our behaviour to keep safe. We also rely on others to manage the locations to reduce the risks. In the town centre we expect those who plan roads to build in safe crossing points. Along the coast we expect landowners to put up warnings or barriers to stop us entering danger zones.

In a city centre we could investigate risks and ways of mitigating risks by:

- using traffic and pedestrian flow surveys to identify three of the most dangerous crossing points. You would need to decide on what times of the day to carry out the surveys and which days of the week;
- asking people what they think about the management of pedestrian and traffic flows using a Likert Survey. Use strong statements like '*The best way to reduce the risk of accidents is to pedestrianise the whole of the town centre*' to see if the views of pedestrians are different from those of motorists;
- using a bipolar recording sheet to record observed behaviour of pedestrians. Pedestrians could be placed into categories (for example, young children, the elderly, those with disabilities) to see if different groups used Pelican Crossings in different ways to minimise risks.

Figure 1 Separating busy traffic from pedestrians in Ipswich

Section 3

Figure 2 Eroding cliffs at Happisburgh, Norfolk

In a coastal area we could investigate risks and ways of mitigating risks by:

- carrying out a perception survey to find out why caravan owners are prepared to risk locating their holiday homes near an eroding coastline. Do the benefits outweigh the risks? Have the risks grown in recent years?
- mapping a stretch of coastline to see how risks have been managed by slowing rates of erosion through sea defences. Why are some areas left without sea defences? Are some defences better than others?
- using transects to survey a section of coastline to see if the current risks are the same or different from the evidence found on historic maps and aerial photos;
- using fieldwork along with Ordnance Survey maps to identify how the loss of coastal roads can be disruptive to those living and travelling along the coast.

Activities

1 **Study Figures 1 and 2. For each photograph, suggest one other way that risk or mitigating risk could be investigated.**

2 **Think about each of the following environments:**
   - an area surrounding a large sports stadium or leisure complex;
   - a brownfield site which is being considered for a major housing development;
   - a housing estate where crime rates (for example, burglary rates, street crime, car theft) are rising.

   a) Suggest how you might investigate the mitigation of risk in each environment.

   b) List the difficulties of investigating risk in each environment.

3 **Select one other environment (not listed above).**

   a) In pairs, share your ideas about how risk and the mitigation of risk could be investigated in the environments chosen.

   b) Give each of your investigations a clear aim.

   c) Suggest at least two suitable fieldwork methodologies (techniques) to collect primary data.

   d) Identify a suitable source of secondary data that could provide useful evidence to support your chosen investigation.

Flood risk and reducing (mitigating) flood risk

https://www.gov.uk/check-flood-risk

The risk of river flooding and the actions people take to mitigate against the risk is an obvious area of geography that lends itself to fieldwork investigation. Clearly it is not safe to undertake fieldwork when the flood is actually occurring. However, there are a range of fieldwork enquiries that could be undertaken in areas prone to flooding. The Environment Agency (EA) provides a useful tool for selecting potential places to collect your data:

Step One Follow the link on the website to 'view map'.

Step Two Enter a postcode for a street somewhere within your target area. Figure 3 is an example of what you will see. Select two locations if you want to do a comparative study.

Step Three Use the zoom feature to gain more detail and see the map at a different scale.

Step Four Do a web search to see if previous floods in the area have been reported in newspapers and in video footage.

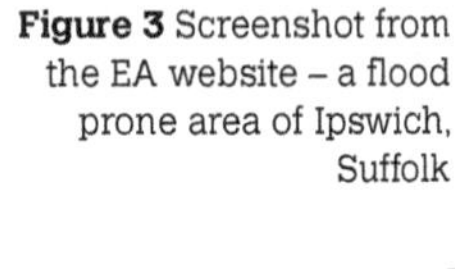

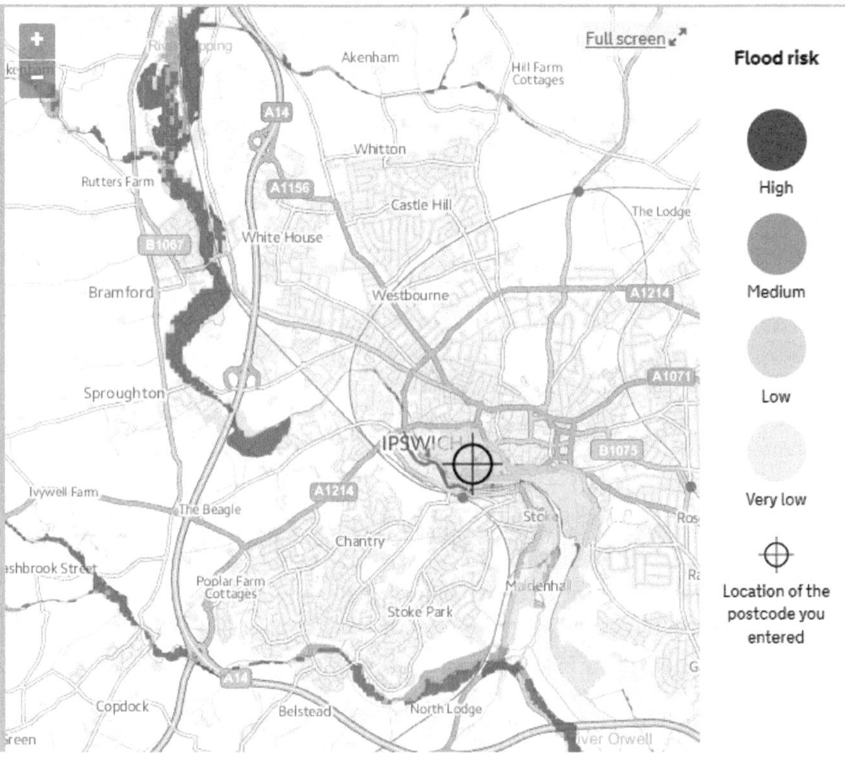

Figure 3 Screenshot from the EA website – a flood prone area of Ipswich, Suffolk

Activities

1 a) Identify an enquiry question about flood risk near to a river.
  b) Identify the fieldwork methodologies you want to make use of in the investigation. Use the guidance below to help you.
  c) Evaluate the strengths and weaknesses of carrying out the investigation as an individual compared with working as part of a group where you work as a team to collect and then share the data.

How to collect evidence for this investigation

- Measure flows such as infiltration in different land-use areas within the target zone.
- Use transects at right angles to the river:
  - to survey the land use. This could be mapped across the flood plain, by dividing up the land into 50m x 50m squares and recording the dominant land use in each square;
  - to show land-use vulnerability by estimating the potential economic value (use simple categories);
  - to show vulnerability in terms of height above normal river levels.
- Use questionnaires to determine different perceptions of the risk involved in living and working in a flood risk area (see below). Perceptions and concerns could also be investigated in relation to the potential management of a flood event.
- Map the transport infrastructure for the wider area to identify the likely impacts on traffic flows should a flood occur.
- Take digital photographs, or draw field sketches, of any flood reduction measures that have been built. Annotate these to show (in one colour) how the flood reduction methods are expected to work. In another colour, annotate any disadvantages these measures create.

Perception Surveys

Some risks we experience first-hand, such as the risk of crossing a busy road. Some risks we will be aware of because we can see clues such as CCTV cameras and neighbourhood watch posters. It suggests an area is at risk of crime or vandalism. When floods threaten or when people are informed about the high cost of flood mitigation, people have strong views. As well as asking people directly threatened by the flood, it is also worth asking people who are not directly threatened by such a danger. They may also have viewpoints because they are tax payers (so provide the money for the mitigation schemes), or they may use the area, perhaps for leisure.

In 2017 a student in Ipswich interviewed a resident who lives in a high risk flood area. The student simply asked the resident an open question and recorded the response on their mobile phone (see Figure 4).

Figure 4 The response of a resident in Ipswich to an open question about flood risk

'My flat is on the second floor so most of my possessions were not damaged in the last bad flood in 1977. My car was damaged beyond repair though. The people on the ground floor lost most of their furniture and many of their personal possessions. We always said that the council should have done more to prevent the flood but it wasn't until 1980 that work started. The downside of the flood defences is that the walk along the river front has been spoilt by the flood wall. We have not been affected since then so I feel safe now. The newspaper has reported that more money needs to be spent on improving the defences in Ipswich because of rising river levels and higher tides. They said we should also expect more stormy days. On my pension, I am not sure how I will be able to afford higher taxes to pay for this.'

Activities

2 a) Use Figure 4 to identify the strengths and weaknesses of asking open-ended questions when investigating perception of the flood hazard.

b) List three essential things you need to do before you carry out an open-ended interview like this. Share your list with a partner and merge the two.

3 Use the guidance provided on pages 76-77 to construct a perception questionnaire to find out about:

a) different people's perception of the flood hazard;

b) different people's perception of the ways that the flood risk has or can be reduced (mitigated).

Time to practise questions about fieldwork on mitigating risk

Ten students decided to investigate the risk of flooding along the coast. They focused their fieldwork on five sites along the Suffolk coast. Two of the sites can be seen on the Ordnance Survey extracts in Figure 5. The scale of the map they used was 1:25 000.

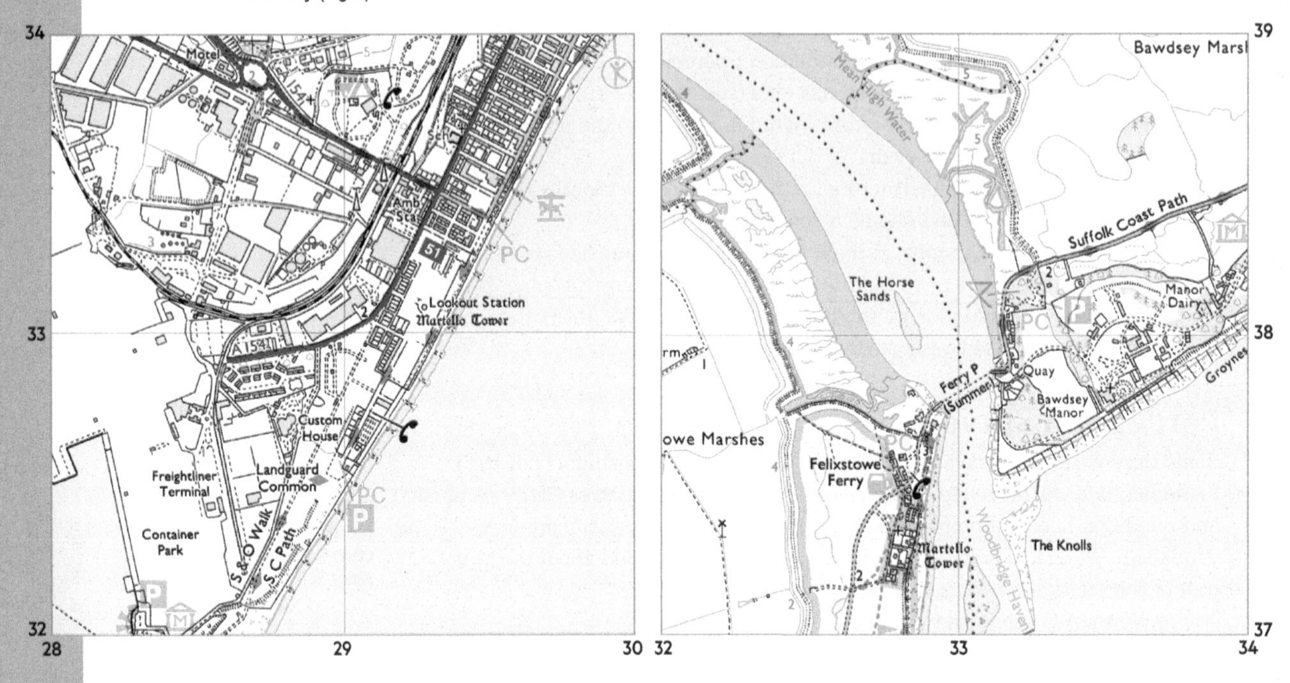

Figure 5 The location of the investigation in southern Felixstowe (left). The location of the investigation in Felixstowe Ferry and Bawdsey (right)

1 a) In the planning stage of their fieldwork they decided to use Ordnance Survey maps to locate the 5 survey points. Give **two** reasons to explain why it is important to plan the exact locations for the survey points. [4]

b) Explain why it may be better to use a 1: 25 000 rather than a 1: 50 000 map for planning your fieldwork. [2]

c) The group decided to identify some of the key stakeholders who would have a view on managing the flood risk at each location. Use the maps to suggest four stakeholders that they should try to interview. [4]

2 **As part of their investigation they wanted to assess how people felt about the possible risk of coastal flooding at each of the five locations. They created a simple survey sheet to do this. They asked 100 people the following question at each of the five locations:**

'Do you think that climate change will increase the risk of coastal flooding at this location in the future?'.

They asked each person to tick one of three boxes:

No increased risk ☐

Some increased risk ☐

Serious increased risk ☐

The results from four of the five survey locations were plotted on a triangular graph (Figure 6).

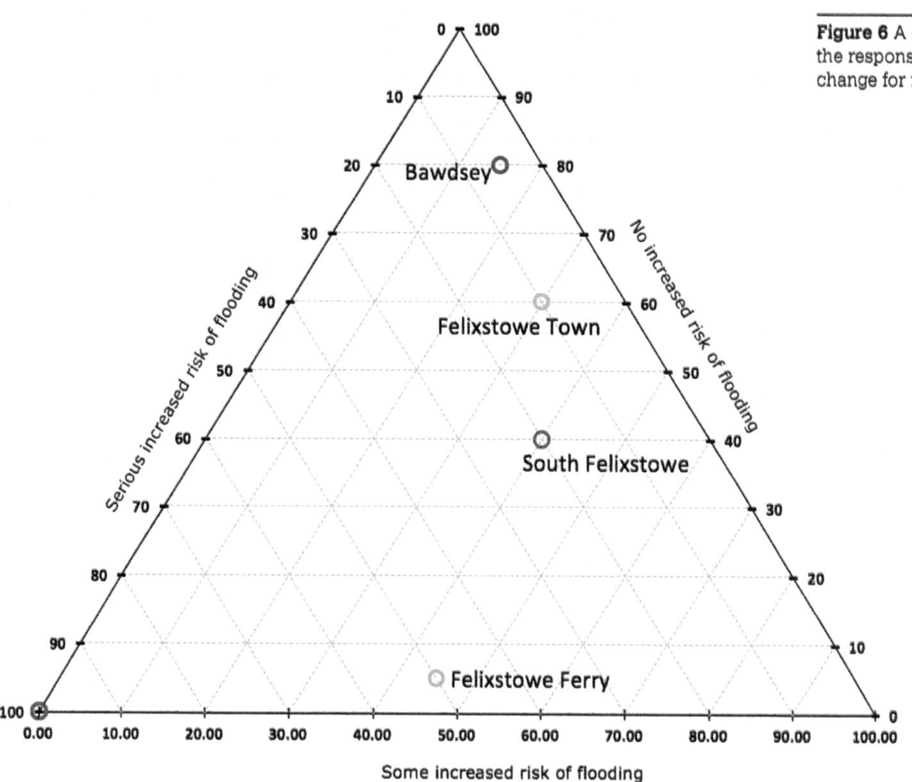

Figure 6 A graph to show the responses about climate change for four locations

a) Use Figure 6 to state how people responded to the question at the Felixstowe Ferry location. [3]

The results for Shingle Street (the fifth location) are missing from the graph. These results are shown in the table below. [3]

Location: Shingle Street		
No Increased risk	Some increased risk	Serious increased risk
20	30	50

b) Ask your teacher for a blank triangular graph, or download one from the internet. Plot the results for Shingle Street on the graph. [3]

3 a) Study the results for South Felixstowe and Felixstowe Ferry on Figure 6. Use evidence from Figure 5 to suggest reasons why the responses at these two locations are so different. [6]

b) Give one human and one physical reason to suggest why the results for Felixstowe Town and Bawdsey show that people think the risk from climate change is lower here. [4]

4 a) In the evaluation of their work the students recognised that most of the people surveyed were visitors, not local people. Does this make the investigation less valuable? Yes or No? Explain your choice. [6]

b) What would be the value in repeating this enquiry in ten years time? [6]

Chapter 15

The concept of Sustainability

Sustainability is a measure of how well a place or environment is coping with social, economic, or environmental change.

Learning objectives

- Understanding the concept of sustainability and how it might be investigated through fieldwork.

What is this concept about?

Investigating sustainability through fieldwork is a useful way of making connections between human and physical geography. It is a chance to consider what is happening now and also to look into the future. As a concept it can be studied at a global scale, when geographers consider the future of the planet, or at a local scale. For example:

- how we can plan new and existing sustainable communities at a time when there is demand for our towns and cities to grow;
- how we manage the coastline at a time when sea levels are rising and wild weather events cause more frequent storms threatening flooding and erosion. Is it sustainable for some communities to remain in these coastal locations?
- why roads and railways are under so much pressure and how we could change our behaviours to make travel more sustainable;
- how people can make full use of the landscape for recreation and enjoyment without harming the environment, or disrupting the people who live there.

Each of these issues creates debate about a sustainable future for the area where we live. Fieldwork allows us to investigate these issues first-hand.

Figure 1 Is tourism at this honeypot sustainable? Dovedale in the Peak District. Over 3000 people regularly visit this place each Sunday during the summer

We could investigate sustainable tourism in a tourist honeypot such as Dovedale by:

- investigating the effectiveness of signage and information boards. The location of signs could be compared to flows of walkers to see whether or not they are located in the most useful places. You could use a bipolar survey to investigate whether the signs are regarded as useful by visitors;
- measuring the percentage distribution of plants across footpaths to investigate the impact of trampling on sensitive vegetation and whether trampling is managed effectively.

Figure 2 Villages across the UK are changing. Many are growing, others are in decline. How sustainable is our village? Orford, Suffolk

We could investigate the sustainability of a rural community by:

- surveying the availability, or access to, local services such as the availability of libraries, post offices, pubs, doctors, grocery stores;
- using questionnaires to collect data about different aspects of sustainable living from different stakeholder groups. For example, views on community cohesion and involvement, inclusiveness, building design, safety, perception of crime / traffic / travel options, pollution;
- creating connectivity maps showing time and distance to analyse how easy it is for villagers to access nearby places for work, school, leisure, health.

Activities

1 a) Study Figures 1 and 2. For each photograph, suggest one other way that sustainability could be investigated through fieldwork.
b) Share your ideas and create a list of ten possible fieldwork investigations.
c) Order the list in terms of difficulty of collecting field evidence. Give reasons to explain why some field investigations are more difficult than others.

2 **For each of the following, suggest how you might use fieldwork to investigate sustainability. Remember, your investigation needs to consider the current situation and look ahead to the future.**
a) A river management scheme.
b) Using brownfield sites for a new housing or retail development.

Do you live in a sustainable community?

People living in a sustainable community have access to the services they need, feel safe and secure, and enjoy a reasonable quality of environment. These features are summarised by Egan's Wheel in Figure 3. Most communities, whether they are in rural or urban environments, have some of the features shown in Figure 3. However, you might want to investigate:

- why one neighbourhood might seem to be more sustainable than other;
- how local residents think that a community could be made more sustainable.

Figure 3 Egan's Wheel

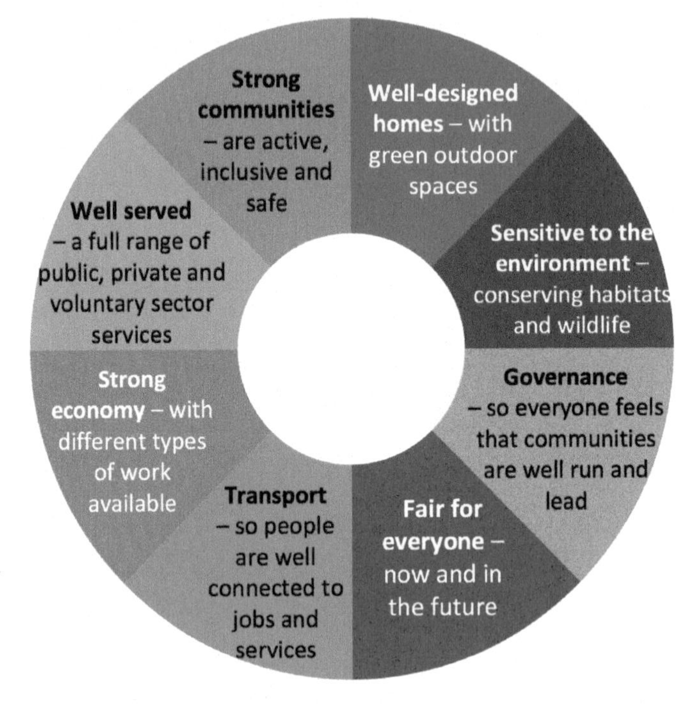

Figure 4 A mobile library provides a service once a fortnight in this remote rural community in mid-Wales

We could investigate whether a community is sustainable by:

- mapping the location of green spaces including parks, recreational areas and churchyards. Interviewing people about how they use these spaces.
- monitoring traffic and taking noise readings to identify dangerous, busy or noisy locations;
- auditing community services such as the location of post offices, health centres, libraries, public transport, or sport/leisure facilities. Asking local people whether they value these services and investigating whether they are accessible to people who are partially sighted and those in wheel chairs.

Investigating sustainable coastal communities

The risk of flooding in some coastal communities is particularly high. In other coastal areas, communities are at risk from erosion. We could investigate whether a coastal community has a sustainable future by:

- investigating the quality of coastal defences;
- considering the possible economic, environmental, and social impacts of a flood;
- using Egan's wheel to help construct a questionnaire. Use it to find out how local residents perceive the sustainability of their coastal community.

Coastal flooding poses a much bigger risk in some coastal areas than others. However, coastal defences are expensive and not all coastlines are worth protecting. We could investigate the social, environmental and economic impacts of a coastal flood by using a weighted Environmental Quality Index Survey (EQI) like the one shown in Figure 5.

Step One Design criteria for measuring the environmental, economic, and social impacts of a coastal flood or coastal erosion.

Step Two Discuss whether you think each of the criteria in your EQI is of equal importance. You could do this in class. Alternatively, you could use a pilot survey to get the opinion of the public.

Step Three Weight each criterion based on your findings in Step Two.

Economic	Businesses at risk	Large shops or factories at risk	30	Small shops at risk	15	No businesses at risk	0
	Transport links at risk	Major roads or rail lines at risk	20	Minor roads at risk	10	No roads at risk	0
Environmental	Damage to habitats	Rare habitats at risk	10	Mixture of habitats at risk	5	Some low value habitats at risk	0
	Damage to buildings	Historic buildings at risk	10	Ordinary buildings at risk	5	No buildings at risk	0
Social	Risk to community	Vital services at risk	20	Non-vital services at risk	10	No services at risk	0
	Risk to residents	Large population including elderly	30	Small population	15	No local residents	0

Figure 5 A weighted EQI to help assess sustainability of a coastal community

Activities

1. **In groups, discuss the weighting factor used by the students. Is a weighting factor necessary?**
2. **What secondary data could be collected to investigate the likelihood of coastal flooding before the fieldtrip?**
3. **Describe how you would show the EQI scores for several locations along a stretch of coast. For students exploring sustainable management options, how useful would this map be?**

Time to practise questions about fieldwork on sustainability

A school decided to investigate how far the market town of Hadleigh, Suffolk, could be considered a sustainable town.

One geography group of twenty students decided to divide the town into two sectors. Ten students worked in the northern sector and ten in the southern sector. Each student used questionnaires and a bipolar survey sheet. They collected their data for half a day. After the fieldwork, they exchanged the data collected in each of the two sectors.

Another group of ten students decided to use the same data collection sheets but to survey the whole town in half a day.

1 a) Identify two advantages and two disadvantages of dividing the town into two sectors and then exchanging the data collected. [4]

b) Suggest one way to minimise any disadvantages arising from shared data collection. [2]

c) Investigations to find out how sustainable towns like Hadleigh might require students to survey a large area before they reach their conclusions. Explain why. [3]

2 A range of bipolar survey sheets have been developed to support investigations into the sustainability of places. Five of the eight segments of the Egan Wheel have been used by the students to design bipolar statements. They did not use three segments:

- **Fair for everyone**
- **Governance**
- **Strong communities**

a) Select two of the three segments above. Explain why each one is difficult to include as part of a fieldwork investigation. [4]

b) For the remaining five segments of the Egan Wheel, a bipolar sheet was created. For each segment, pairs of opposing statements were designed to allow students to focus on different aspects of the segment. Two segments are shown in Figure 6. For each segment, suggest a second pair of statements to widen the scope of the investigation [2]

Figure 6 Extract from a bipolar survey sheet

Segment of the Egan wheel	**Negative statement**	**Score**					**Positive statement**
		1	**2**	**3**	**4**	**5**	
Transport	**There are no public transport links from the place**						**There are very good public transport links from the place**
	Suggest a second						*Suggest a second*
Economy	**There are no jobs available within the local area**						**There are a wide range of jobs available within the local area**
	Suggest a second						*Suggest a second*

3 The ten students surveying the whole of Hadleigh shared their data. These were collated onto the sheet below (Figure 7). A mean score for each student has been calculated. A mean score for each of the five segments of the Egan Wheel has also been calculated.

Student	Transport	Environment	Economic	Housing	Services	Mean Score for each student across the 5 segments
1	5	7	5	8	7	6.4
2	5	7	4	9	7	6.4
3	4	7	4	8	8	6.2
4	4	7	4	7	8	6.0
5	3	7	5	8	6	5.8
6	4	7	4	8	6	5.8
7	4	8	5	8	6	6.2
8	4	7	6	8	7	6.4
9	**2**	**3**	**2**	**3**	**3**	**2.6**
10	5	6	5	8	7	6.2
Mean score for each segment						The mean of means = 5.8
	4.0	**6.6**	**4.4**	**7.5**	**6.5**	

Figure 7 The collated results of ten students.

a) What conclusions can you reach about the degree to which Hadleigh could be considered a sustainable town? Use data from the table to support your answer. [5]

b) Study the data provided by student 9 in the table above. Describe how the data is similar to and different from the data from the rest of the students who used the survey sheet.
[4]

c) In cases like this, when the data set is not symmetrical, it is sometimes better to use the median value. Explain why. [2]

d) Calculate the overall median score for the group of students. [1]

Chapter 16

The concept of Inequality

Features and resources are rarely distributed evenly through the environment. Inequality occurs where one place, or one group of people, has more than its fair share of something.

Learning objectives

- Understanding the concept of inequality and how it might be investigated through fieldwork.

What is this concept about?

Geographers are interested in inequality at different scales. Within the UK we can investigate inequality comparing one region with another, perhaps looking at health care or education provision. At a smaller scale, particularly when we want to collect primary data, we can focus our fieldwork on very specific issues within one town or one small rural area. This might include comparing where we live with other locations in terms of, for example, access to jobs, schools, safe parking, or affordable housing. An even tighter focus could look at inequality for a particular group such as the teenagers or those with disabilities. In this case you might investigate how the chosen group might be disadvantaged in terms of access to leisure facilities or ease of movement around a shopping area. Fieldwork investigating inequality is generally focused on issues linked with human geography. Investigations that compare one location with another are an ideal starting point when you are deciding on your hypothesis or enquiry question.

Figure 1 Identifying suitable enquiry questions for a fieldwork investigation on inequality

Activities

1 **For each of the enquiry questions shown on Figure 1 suggest a SMART hypothesis that could be tested. Use the guidance on pages 8-9 of this book.**

Quality of Life Surveys

Comparing quality of life in two locations is one of the most popular fieldwork investigations when studying inequality. The collection of primary data for such investigations allows you to use many of the methodologies and techniques explored in Sections 1 and 2 of this book. You can also support your fieldwork by accessing a wide range of secondary data sources. You should make sure that your data gathering includes social, economic and environmental features as they all influence quality of life.

Avoid creating survey sheets that assess a limited range of features. You should try to include qualitative surveys that assess the features in Figure 2.

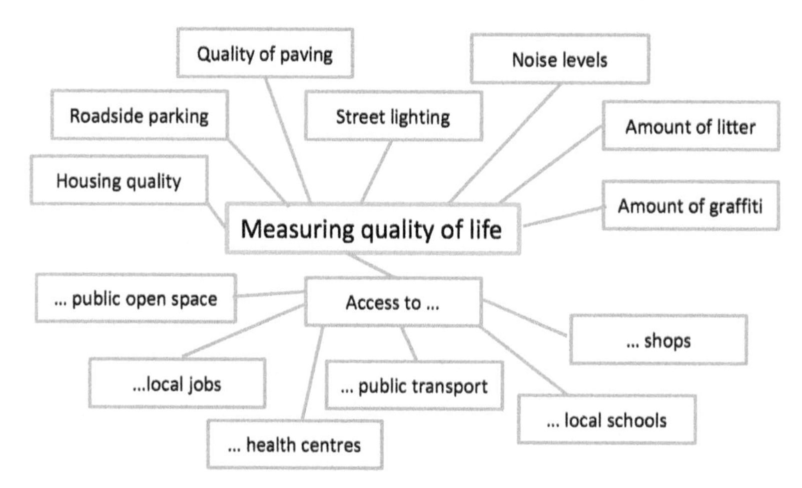

Figure 2 Features that influence quality of life

Why do you need to use secondary data?

It is not possible to get primary data on some of the most important factors that influence quality of life through fieldwork. You need to combine your primary data with selected secondary source material. Pages 26-27 outline the valuable data that you can collect from the National Census, information such as:

- unemployment rates;
- the number of hours people work;
- health of the population;
- levels of car ownership.

This secondary data can provide evidence needed to build up a better understanding of quality of life in an area. If you are doing a comparison study, this information is available at ward level, statistics for each neighbourhood can be accessed by using the postcode.

As well as the National Census, other websites provide valuable additional information to support quality of life surveys. Use the web links listed below.
www.traveline.info
www.police.uk
https://www.google.co.uk/intl/en_uk/earth/

Activities

2 Study Figure 2. For each leg of the spider diagram suggest a bipolar statement that could be added to a survey sheet to assess quality of life.

3 a) In small groups discuss why the four items suggested from the National Census would be valuable when investigating quality of life.

b) If time prevented you from exploring all four of the items, which two would you choose to support your primary data gathering? Justify your choice.

4 Work together in groups of three. Each of you should review one of the data websites. Explain to your partners how your link would help in your quality of life investigation.

Access to services

The concept of inequality provides many opportunities for investigating access to services. You could investigate access to transport, leisure facilities, or shops. In fact, almost any service or range of services could be the focal point of the study. Some groups of people appear to have better access to some services than others. Your enquiry question could reflect this. For example:

- *Do teenagers living in a city have better access to sports facilities than teenagers living in villages?*
- *Do able bodied people have better access to shops than people in wheelchairs or parents pushing buggies?*

You should start by deciding on a group (or groups) of people for your investigation. In the example below the service to be investigated was rural bus services. The target group for the investigation were students aged 16-19 who wanted to continue their education after completing their GCSEs.

Figure 3 is a sketch map which shows the area of study. In this case, the students live in one of two villages outside the market town of Bury St Edmunds, Suffolk. From the bus station in the centre of the town, two schools offering post-16 courses and one college are all within easy walking distance.

Figure 3 Sketch map showing the location of the field investigation

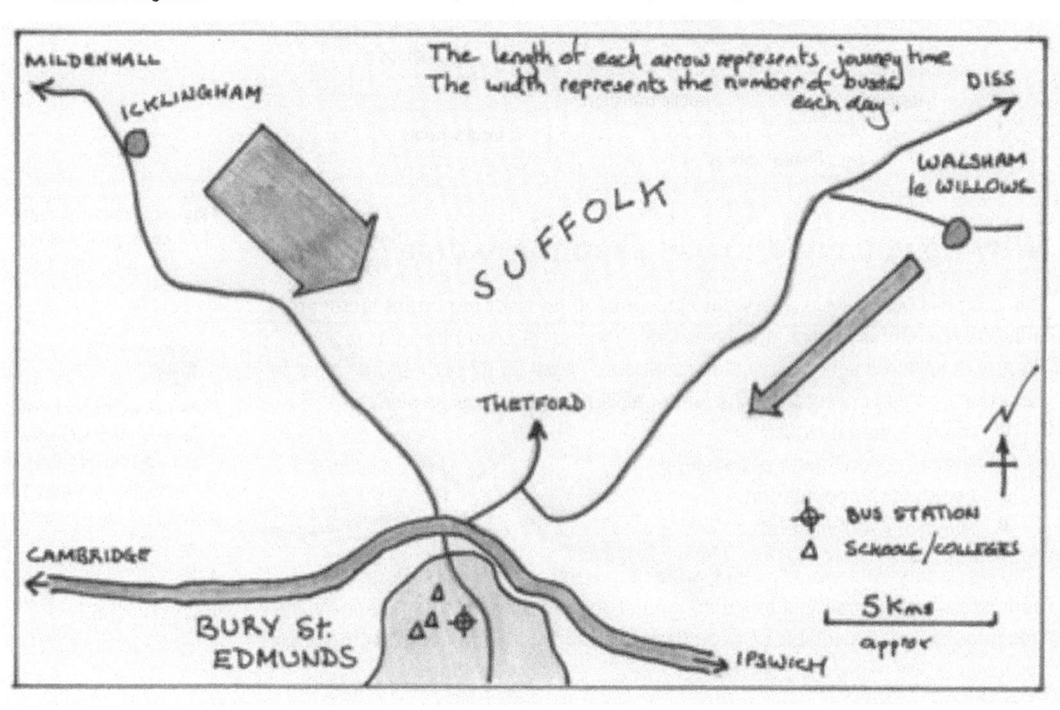

The starting point for the investigation is to use secondary data. Bus timetables can be accessed through the timetable published at the bus stop, or via the internet, to calculate the Bus Coverage Index. This calculation allows you to compare data about public bus services from each village. The Bus Coverage Index is a measure of accessibility. Follow the steps below.

Step One Use the bus timetable to calculate the average journey time from the village to the bus station.

Step Two Add up the number of buses for the selected route (inward and outward) each day.

Step Three Divide the average journey time by the number of buses per day.

Step Four Compare the two scores. The lower the Bus Coverage Index the greater the accessibility.

For the investigation in Suffolk, the scores were calculated as follows.

Icklingham to Bury:
Journey time (28 minutes) ÷ Number of buses per day (26) = 1.08

Walsham le Willows:
Journey time (38 minutes) ÷ Number of buses per day (6) = 6.33

Figure 3 also shows the same data represented graphically. This is a development from the ideas shown on page 40. This shows that this investigation could also be used to investigate geographical flows as well as inequality.

Once the Bus Coverage Index has been calculated, the follow-on fieldwork needs to investigate a wide range of other factors that must be taken into account to fully understand inequality in bus service provision. Visits to each location will enable you to collect the data. Questionnaires could be created to ask the students for their own views.

We could investigate inequality of service provision by:

- asking the students who use the service to complete a Likert survey which asks questions focused on the reliability of the service, how crowded the bus is, how access friendly is it for those with limited mobility, how far the bus timetable matches with the timings of college commitments;
- using secondary evidence from public service timetables to work out how effective the bus service is in linking with other transport hubs. Connection times and time distance lines could be mapped.

Activities

1 Suggest how the sketch map could be improved to show the information more clearly.

2 Although the Bus Coverage Index shows some valuable information for comparing access to bus services, there are some limitations. Identify two limitations about bus services that are not considered in the calculation.

3 Study Figure 4.
Create a questionnaire using a combination of closed and open questions that could be used to investigate access to services in this rural location.

4 a) Name two techniques you would use to collect data in each of the villages. For each technique describe how you would record the data collected.

b) Identify two problems that you might encounter when carrying out the fieldwork. Suggest how early planning would limit these problems.

Time to practise questions about fieldwork on inequality

Students decided to investigate the variation in quality of life in Ipswich by comparing two neighbouring wards. The Gainsborough Ward is dominated by local authority housing built in the 1950s. The Nacton Ward is a mix of suburban housing and urban-rural fringe villages.

Field sketches were used to highlight positive and negative aspects of quality of life in each area.

Figure 5 Two student field sketches showing an example of suburban house in Nacton Ward, Ipswich

A – Private drive – Positive as cars are safer off the road and in a garage.

B – Single yellow lines with notice that states no parking 8.00am – 6.00pm – Negative as friend cannot park close by in the daytime when the drive is being used.

C – Dropped kerb – Positive as car will not be damaged on entry.

D – Wooden blocks stop cars parking on the grass verge. – Positive as vegetation is not ruined.

E – Street lighting – Positive as streets are safer when lit. Negative if light pollution affects bedroom.

F – Incinerator chimney from hospital just behind the house – Negative as house price will be lower with fear of pollution.

G – Front gardens large enough for small trees. Positive as trees increase privacy.

Field sketch A

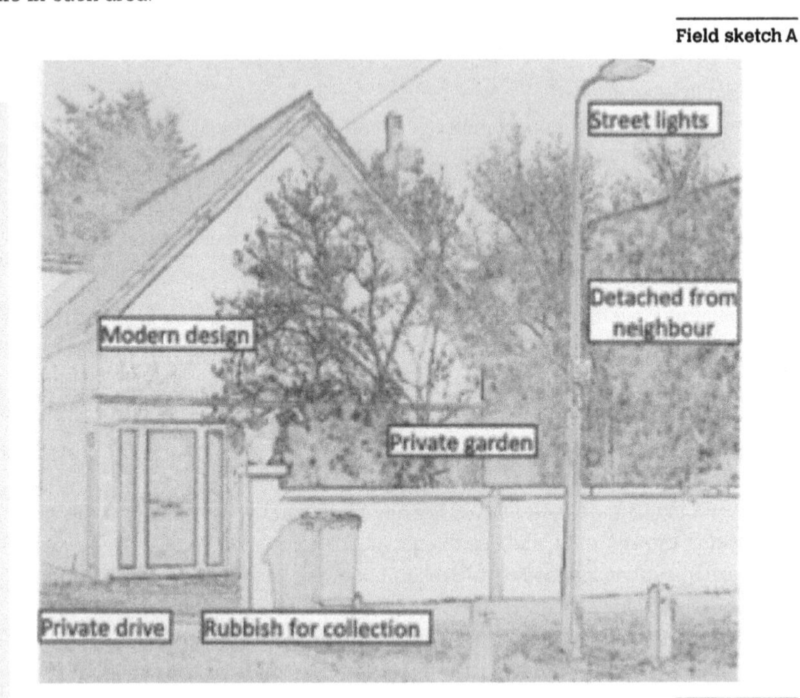

Field sketch B

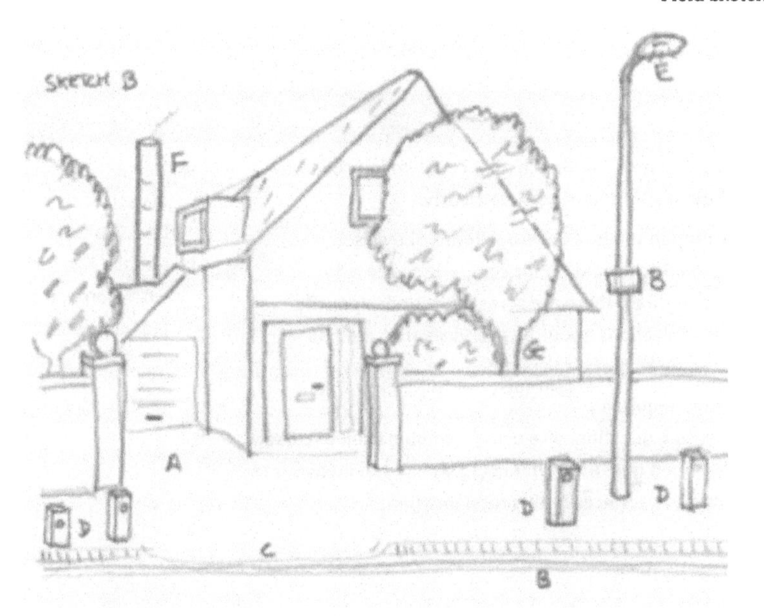

1 **Give three reasons why field sketch B is more useful than field sketch A to support this investigation.** **[6]**

2 **The students used a Likert scale to find out the perceptions of local people in each area. The students didn't feel that they could ask the residents some important things that affect quality of life. They used data from the 2011 National Census (Neighbourhood Statistics by Ward) instead. The students converted all the raw data into percentages.**

Figure 6 Comparative figures for four factors that each affect quality of life. Source: National Census (2011)

Factors that affect quality of life		Nacton Ward	Gainsborough Ward
General health of the population.	Percentage In good health.	75	65
	Percentage In fairly good health.	20	25
	Percentage not in good health.	5	10
Percentage of households with no access to a car or van.		6	29
Percentage of households with no central heating.		2	5
Percentage of adults who do not own a passport.		12	34

a) Explain why the students converted the raw data into percentages. [3]

b) Select **two** of the chosen factors. Explain how they impact on quality of life. [6]

c) Which of the four factors chosen by the students do you think best supports an investigation into comparative quality of life? Explain your choice. [4]

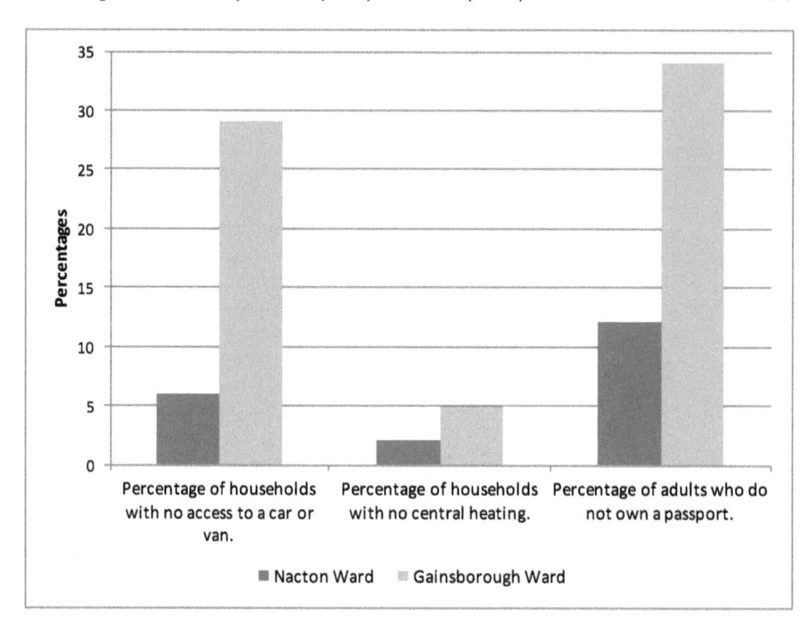

Figure 7 Bar chart representing data from Figure 6

3 **Figure 7 shows how the students used excel to graph some of the information. Suggest an alternative way of representing the data for all four of the factors. Explain why your choice is suitable.** **[6]**

Glossary

Accuracy How close a measurement is to the true value of the object that is being measured.
Aim What you hope to achieve or prove through fieldwork.
Altitude The height above sea level.
Analysis A process in which you make sense of the evidence.
Anemometer A piece of fieldwork equipment used to measure wind speed.
AONB Area of Outstanding Natural Beauty. Countryside in the UK that is protected and conserved.
Axis/y/x The x-axis is the horizontal base line of a graph. The y-axis is the vertical base line.
Bar chart A way to represent data. Values are shown using vertical columns or horizontal bars.
Beach profile A cross section showing the landscape of a beach, both above the water and below it.
Belt transect (continuous & interrupted) A sampling strategy. Data is collected along a line (or transect) usually by using a quadrat.
Bias A tendency to be positive or negative about something or someone.
Bipolar survey A data collection method in which people express an opinion by choosing from opposite pairs of statements.
Branding A geographical concept. How places create an image through redevelopment and advertising.
Causality The relationship between something that happens or exists and the thing that makes it happen.
CBD Central Business District – the town or city centre.
Characteristics Elements or qualities belonging to a place or feature which help define its identity.
Climate change A large-scale, long-term change in the Earth's weather patterns or average temperatures.
Clinometer A piece of fieldwork equipment used to measure gradient / slope angle.
Clone town A geographical concept. How towns/cities have lost their local identity because most of the locally owned shops and businesses have been replaced by national chain stores.
Cluster(ed) A group of similar things gathered or occurring closely together.
Commuting The movement of people between their home and work place.
Concept A big idea that helps us make sense of geography.
Conceptual framework A geographical concept (or big idea) that gives your fieldwork a purpose.
Conclusion A summary statement or judgement that summarises the findings of a fieldwork enquiry.
Connection The link that exists between one set of data and another.
Correlation The connection that links two sets of data (or variables) together.
- **Negative correlation** Values in one set of data increase as values in the other set of data decrease.
- **Positive correlation** Values in both sets of data increase at the same time.

Counting To record how many whole units there are.
Cross section Represents the shape or profile of a feature using measurements of the distances and depths from a horizontal line.
Cycles A geographical concept. How some processes are connected to one another through a system of flows and cycles. The water cycle is a good example.
Data Facts and statistics collected together, for example, as evidence and for analysis.
- **Big data** A very large and complex set of data.
- **Bivariate data** Two sets of data that are connected in some way.
- **Census data** Data about the whole population which is collected every 10 years.
- **Continuous data** Values that can be measured and recorded to one or more decimal point.
- **Discrete data** Values that can be counted and recorded as whole numbers.
- **Primary data** Data that is collected first hand.
- **Secondary data** Data that has already been published in another source.
- **Qualitative data** Evidence that is collected as words, opinions, or images.
- **Quantitative data** Evidence that is collected as number values.

Dataset A group of values (set of data) that is often presented in a table.
Dispersion graph A way to represent data. Data is plotted to show the range of values.
Egan's Wheel A way of presenting the characteristics of sustainable communities named after Sir John Egan.
Enquiry process The process by which a student collects and analyses evidence in order to make a decision or prove/disprove an aim.
Environment The surroundings or conditions in which a person, animal, or plant exists.
EQI (Environmental Quality Index) A technique which uses detailed criteria (statements) to assess the quality of our surroundings.
Erosion A process by which the land is worn away.
Evaluation A process in which you weigh up strengths and limitations. Evaluation is an important stage in your fieldwork enquiry in which you assess how well your fieldwork has gone. For example, you can evaluate the methods you used to collect or represent data.
Evidence Information that can be used to investigate an aim, test a hypothesis, or support an argument. Evidence may include quantitative or qualitative data.
Extrapolation A process by which a graph is used to find a data value that lies outside (or beyond) other data values.
Feature A natural or built part of the environment.
Fieldwork Practical work undertaken in natural and urban environments to investigate geographical questions and hypotheses.
Flow line map A way to represent data. The amount of a flow (such as traffic or pedestrians) is shown using a proportional arrow.
Flows Movement of things through an environment, for example, water, people, money.
Frequency The rate at which something happens over a particular period of time.
Geobrowser A technology which provides a three-dimensional, computer-generated representation of the Earth, or another world, letting the user browse content visually and view different aspects of the environment by rotating it, changing the angle, as if flying over an area.
Geological map A way to represent data. Rock types are shown on a map by using different colours.
Geographical concepts Big ideas that help us understand how the world works.
Google street view A feature in Google Maps and Google Earth that provides panoramic views from positions along many streets in the world.
Grid reference A reference which describes a location on a map via a series of vertical and horizontal grid lines identified by numbers and letters.
Honeypot A geographical concept. A place that attracts many visitors, like bees around a honeypot.
Hypothesis A statement which can be proven to be correct or incorrect based on the evidence collected in your fieldwork.
Identity A geographical concept. The features that make a place special or unique.

Inequality A geographical concept. The ways in which different groups of people have uneven access to such things as jobs, wealth, housing, or services.

Infiltration (rates) The speed at which rain water soaks into the soil.

Interpolate/interpolation A process by which a graph is used to find a data value that lies between other data values.

Interquartile range The middle 50% of a data set. The range from Q1-Q3.

Investigation Using enquiry skills to test a hypothesis or answer a question by collecting and analysing data and evidence.

Irregular Not regular. Does not fit a pattern or set of rules.

Isoline map A way to represent data. Data points with the same value are joined by a line.

Judgement A decision that you must support by use of evidence.

Key The legend of a map or graph that describes the meaning of colours or symbols that have been used.

Kite diagram A way to represent data.

Landform A natural feature in the landscape.

Line graph A way to represent data. Values are shown by drawing a line that joins the data together.

Linear Arranged along a straight or nearly straight line.

Likert survey A data collection method in which people express an opinion by stating how strongly they agree or disagree with statements.

Logarithmic A scale on a graph where the numbers increase more and more rapidly.

Mean A measure of central tendency (or average) in a dataset.

Meanders Sweeping bends that develop as a landform on many rivers.

Measuring/measurement The action of calculating the size, length, distance, area, quantity, depth, range of something.

Median A measure of central tendency (or average) in a dataset.

Methodology A way of collecting data.

Methodological approach A type of investigation that is structured around the way you are collecting the evidence. For example, you could structure your fieldwork around investigating flows or using transects.

Mitigation A geographical concept. The ways in which hazards or risk can be reduced by changing behaviour or managing the environment.

Mode/modal class A measure of central tendency (or average) in a dataset.

NIMBY This stands for Not In My Back Yard. It means that people don't want a particular development (like a new runway) built close to where they live.

Opportunistic A sampling strategy. Data is sampled where it is convenient to do so.

Pattern A regular form or sequence that can be seen.

Perception What someone thinks of a place, idea, situation.

Physical process A series of actions that change a feature over time.

Pictogram A way to represent data. A simple picture is used to show the value of data.

Pilot survey A stage of data collection that allows sampling methods to be tried out and evaluated before a full sample is taken.

Place A geographical concept. The features that give each location a sense that it is different from other locations.

Profile An outline that represents a feature as seen from one side.

- **Beach profile** A diagram that represents the gradients of a beach.
- **Long Profile** A diagram that describes how a feature such as a river varies along its length.

Quadrat A piece of fieldwork equipment. A square frame that is used to sample data (often vegetation).

Quality of life The standard of health, comfort, and happiness experienced by an individual or group.

Rank order A process by which numbers are put into sequence by their value from largest to smallest.

Reliability The degree to which the evidence collected during an investigation can be considered consistent and dependable.

Risk A geographical concept. A possibility or threat of damage, injury, danger.

Rural environment An area of countryside including farms, villages, and smaller towns.

Sample A small part or quantity that can show what the whole of something is like, for example, a group of people drawn from the population of an area used to show what that population is like.

Sampling The process of selecting a sample.

- **Random** Data is sampled at irregular intervals.
- **Regular** Data is sampled at equal intervals.
- **Stratified** Data is sampled in proportion to the size of the original population.
- **Systematic** Data is sampled at regular intervals.
- **Transect** Data is sampled along a line.

Scatter graph A way to represent bivariate data.

Settlement A place where people live.

Sketch A simple drawing that represents the main features of a landscape.

Sketch map A simple plan that represents the main features of a landscape.

SMART A statement, aim, or objective which is Specific, Measurable, Achievable, Realistic, and Timely.

Spatial pattern How features are distributed across a map.

Species A group of living things with similar characteristics.

Sphere of influence A geographical concept. The area around a geographical feature affected by it.

Store(s) A geographical concept. An accumulation or stock of something, for example, an ocean is a store of water.

Sustainability A geographical concept. A measure of how well a place or an environment is coping with social, economic, or environmental change.

Tally mark A quick way to keep track of numbers. A vertical line equals one. Tally marks are made in groups of five. One vertical line for each of the first four numbers and a diagonal line for the fifth.

Text analysis Looking for patterns or trends in words.

Trampling To tread on or crush, for example, plants in the environment.

Transect An imaginary line through the environment. Data and observations can be made along this line.

Trend A change shown on a graph.

Urban environment A human settlement such as a town, city, or area that is made up of buildings, roads and other built features.

Variable A feature, element, or factor which changes.

- **Dependant variable** A variable whose value depends on the value of another factor (such as distance, time, or height).
- **Independent variable** A factor that causes other data to vary.

Vegetation Plants, shrubs, trees.

Velocity The speed at which something flows.

Vulnerability A geographical concept. The extent to which people are unable to resist harm or risk.

Index